P9-AQR-797

STATISTICAL INFERENCE

BASIC CONCEPTS

Richard B. Ellis

Northern Essex Community College
Haverhill, Massachusetts

PRENTICE-HALL, INC., Englewood Cliffs, N.J.

Library of Congress Cataloging in Publication Data

ELLIS, RICHARD B.
Statistical inference: basic concepts.

Bibliography: p.
1. Statistics. I. Title.
HA29.E483 519.5 74-6289
ISBN 0-13-844621-0

10 9 8 7 6 5 4 3 2

Printed in the United States of America

PRENTICE-HALL INTERNATIONAL, INC., *London*
PRENTICE-HALL OF AUSTRALIA, PTY. LTD., *Sydney*
PRENTICE-HALL OF CANADA, LTD., *Toronto*
PRENTICE-HALL OF INDIA PRIVATE LIMITED, *New Delhi*
PRENTICE-HALL OF JAPAN, INC., *Tokyo*

ACKNOWLEDGMENTS FOR CHAPTER OPENING QUOTES

CHAPTER 2 M. J. Moroney, *Facts from Figures* (Baltimore: Penguin Books, Inc., 1956), p. 173. By permission of Penguin Books Ltd., London.

CHAPTER 4 Jerome S. Bruner, *The Process of Education* (Cambridge, Mass.: Harvard University Press, 1963), p. 46. By permission of Harvard University Press.

CHAPTER 8 From the book *Freedom and Beyond* by John Holt. Copyright © 1972 by John Holt. Published by E. P. Dutton & Co., Inc., and used with their permission.

CHAPTER 12 Jerome S. Bruner, *The Process of Education* (Cambridge, Mass.: Harvard University Press, 1963), pp. 14 and 64. By permission of Harvard University Press.

CHAPTER 15 M. J. Moroney, *Facts from Figures* (Baltimore: Penguin Books, Inc., 1956), p. 371. By permission of Penguin Books Ltd., London.

CHAPTER 16 James R. Newman, *The World of Mathematics*, Vol. II (New York: Simon & Schuster, Inc., 1956), p. 1170. By permission of Simon & Schuster, Inc., New York, and George Allen & Unwin Ltd., London.

Only God is omniscient.

Contents

Preface

OBJECTIVES

"When one considers the variety and extent of the demands for an 'ideal' first course in statistics, one recognizes the impossibility of having any one course come even close to the ideal" (*Introductory Statistics without Calculus*, by the Committee on the Undergraduate Program in Mathematics, published by the Mathematical Association of America, 1972).

This book is designed to familiarize the student with some of the powerful concepts of statistical inference, their wide application and philosophical implications, and, hopefully, their relevance to the student's own life. An understanding of these basic concepts of inference may best be attained by taking practical problems, asking reasonable statistical questions, applying appropriate solution techniques, and then making sound statistical decisions.

Immediately in mind I have the liberal arts student who is also studying philosophy, history, literature, sociology. . . . I believe statistical theory should not be presented in isolation from the other subjects that the liberal arts student is supposed to be learning about. The typical student has long been subjected to too much of taking courses for their own sake. Educators, if they are to sell education in the future, are going to be forced more and more to show the relationships of the wares they are selling and the relevance of these wares to the student's life, and to life in general outside the college.

Major Objectives

1. To appreciate the story of man's struggle with variability and uncertainty.
2. To see how some basic statistical concepts can enable one to estimate, test hypotheses, establish relationships, and make decisions.

3. To develop an understanding of these concepts by employing appropriate formulas, tables, and calculations.

STRUCTURE

Statistical inference is here pursued in the spirit of inquiry. From organized information we seek ways to estimate, test hypotheses, and check relationships. The struggle to predict from a world of uncertainties provides a sense of drama and accomplishment. We reach a climax in the last chapter when we come up with not only a test of significance but a way of describing a relationship mathematically and actually producing a numerical expression of how good that description is. Correlation is the climax to our drama.

This attempted structure for a one-semester course forces me to cut some corners. Descriptive statistics is not developed any more than the text requires. Probability is developed only insofar as we shall actually use it later on in the book. The necessity for representative sampling is dwelt upon at some length, but the student is referred to other books for various sampling techniques.

The structure is set to the stated objectives. However, the instructor may, and certainly will, make his own sequence changes, assign alternative and supplementary material, and omit some of the material he doesn't deem as important as the author did, in order to accommodate himself, his department, and his students.

STYLE

The book is pretty much my own classroom approach (mistakes not incorporated, I hope). The informal style springs from a desire to talk with the students and work along with them, instead of erecting an imposing, impersonal mathematical structure.

Definitions grow out of the discussion and are not set apart in boldface type. Some of the lessons learned from programmed instruction are employed. Repetition is a deliberate attempt at reinforcement.

The spirit of the style is light at times, in the hope that the student will catch some of the fun I have in class. But all along there is the very serious side, too. The deep significance of the growth of statistical inference in our modern world is emphasized by historical references, literary quotations, special comments, and side discussions.

EXERCISES

The exercises are not diluted for nonmathematical students. Calculating for these concepts, the student doesn't have to know calculus. He needs very little algebra. He does need basic arithmetic, which he is probably weak in. Basic computational arithmetic is given some attention along the way; the attention is generally toward what accuracy and precision it is practical to maintain. The addition, subtraction, multiplication, and division may be done "by hand," though one of the many pocket or desk calculators would greatly reduce the hackwork. The occasional preoccupation with computation, however, is only a means to an end; it is the basic concepts of statistical inference that are of prime importance.

For all of the exercises that require calculation, I have provided, between the appendix and the index, what I hope are reasonable results for the student to compare his own work with.

The exercises begin with several problems that can be handled in the same manner as the examples in the text. Variations on this theme then appear, interspersed with problems that belong to previous chapters. (You can't adequately challenge a student if he can tell the nature of the problem by knowing what chapter he is in.) And there are a few problems that can't be solved at all—usually because insufficient information is given. Recognizing that a solution is impossible requires real understanding, too.

I, and probably you, have students who, after the class discussion, do the exercises without reading the text. I have tried repeatedly here to ask a few questions in the exercises that will encourage the student to see what the text has to say. There are also questions which I think may make a reading of the text more meaningful. So, rather than asking him to turn forward to the exercises and then come back and read the text, I have put the exercises first in each chapter. Some of the students will skip over them, I know. But some may pick up an enlightening glimpse of what is coming. Others may find provocative questions to search out answers for. Thus, the questions in the exercises are designed to introduce the text as well as to look back at it.

Some of the concepts are persistently elusive. At times they seem to appear with great clarity and then strangely slip away. To really capture them, the problem-solving activity must be pursued steadfastly—by doing all of the problems, even throwing solutions away and redoing them, trying different approaches to the same problem, and, especially, making up problems on one's own. A student's success is somewhat proportional to the amount of paper he uses in doing his homework.

I suggest that he be encouraged to ask with every problem something like this:

1. Very specifically what is the question to be asked?
2. Do I have the information necessary to answer this question?
3. What sequence of operations shall I follow and what formulas do I need? (See Table VIII in the appendix.)
4. Does my decision, based upon the results of my calculations, appear to be a reasonable one?

ACKNOWLEDGMENTS

I am indebted to the Literary Executor of the late Sir Ronald A. Fisher, F.R.S., to Dr. Frank Yates, F.R.S., and to Longman Group Ltd., London, for permission to reprint parts of Table IV from their book *Statistical Methods for Research Workers*; also to Professor E. S. Pearson, University College, London, and the Biometrika Trustees for permission to reprint Tables 8 and 18 from their *Biometrika Tables for Statisticians*; and to the Iowa State University Press to use Table A11 from their book *Statistical Methods*. I am grateful also for permissions from authors and publishers to reproduce the many quotations in this text.

I thank Arthur H. Wester, Mathematics Editor of the College Division, Prentice-Hall, Inc., for his encouragement; and I thank Margaret G. McNeily, also of Prentice-Hall, for her competent handling of production matters. I am indebted to Beverly Tarolli for her cheerful devotion to typing and to Barbara Ellis for the strong supporting role she has played during the many months of labor.

RICHARD B. ELLIS

• I •

SAMPLING AND PROBABILITY DISTRIBUTIONS

• 1 •

Our Uncertain World

Happy the man who could search out the causes of things.

VIRGIL (70–19 B.C.)
Georgics, II

I pass with relief from the tossing sea of Cause and Theory to the firm ground of Result and Fact.

WINSTON CHURCHILL
The Malakand Field Force (1898)

OBJECTIVES

To review man's search for causes and to see how, even without the causes, statistical inference gives us a way of estimating, testing hypotheses, and finding correlations.

1–1 EXERCISES

1. Give an example of a happening in your personal experience that you would refer to as pure luck. Try to explain why you call this luck.

2. Is there a difference between luck and chance? What do you mean?

3. What do you mean when you say something happened and it wasn't chance?

4. Maybe you'd better explain what you consider "causal relationship" to mean.

5. Can there be a causal relationship with the cause being just chance? How?

6. Can things happen without a cause? Don't be too sure.

7. Can there be forces at work that are not causal forces?

8. How do you determine which of the forces are the causes?

9. The formula for the distance covered by a body at constant velocity is $d = vt$. Does this indicate a causal relationship? Explain. Is time one of the causes?

10. Do the same causes always produce the same results? Give an example of the same cause producing different results. Of course, it still depends on what you select as the cause.

11. When you explain a happening, does this mean citing causes? How else could you *explain*?

12. Do you need these causal explanations to give you security in our uncertain world? What's your reasoning here?

13. Can there be, then, any assurance in a situation where the circumstances are too complex for us to account for causes?

14. When you can't find a cause or when the situation is too complex to sort out the causes, what basic techniques of statistical inference make it possible for you to estimate, test hypotheses, and find relationships?

15. You are going to have to face challenging problems. Good luck!

1–2 THE SEARCH FOR CAUSES

From the beginning man has sought cause-and-effect relationships. Stick a needle in the tribe leader's image and he dies—isn't this a causal relationship?

In much of man's history, religion has supplied answers—though generally it has been the philosophical ones that still stand: ". . . for whatsoever a man soweth, that shall he also reap."

With people like Galileo (sixteenth century) and Sir Isaac Newton (seventeenth century), men of science began more and more to take the view that every happening had an assignable cause based on "the laws of nature." Newton found that the force necessary to keep the moon in her orbit and the force of gravity on the surface of the earth "answer pretty nearly"; that is, the existence of the gravitational force of the earth comes pretty near to completely explaining the elliptical orbit of the moon. By the twentieth century this faith in the effectiveness of explanation by physical causes had invaded even psychology in the form of behaviorism, human behavior being seen as a system of responses to physical stimuli. Ring the bell and the dog drools—so why not you?

But the laws of nature are beginning to appear as man's oversimplified generalizations. "The volume of a given mass of gas varies inversely as the pressure." And these relationships become probabilities rather than certainties. In one of his most remarkable achievements, the harnessing of nuclear energy, man faced the phenomenon of radioactivity, where alpha particles fly off. He could not say why they fly off. He could not say when they would fly off. He could not say where they would fly off. He could only say it was probable that in a certain time interval, in a certain space interval, and under certain circumstances one *would* fly off! And nuclear power plants are now helping to supply your house with electricity.

So the circumstances turned out to be much more complex and subtle than a simple cause-and-effect explanation had portrayed them to be. We have had to resort to statistical analysis. This approach has become the core of much modern scientific reasoning. We shall see that this statistical treatment amidst life's uncertainties, variables, and dissimilarities has radically changed man's view of his universe.

There is undoubtedly some parallel between the story of man's growing up and the story of your growing up—first blind faith, then causal explanations, then a simple acknowledgment of the existence of certain relationships, whatever the causes. There is a relationship between your illness and the diphtherial bacteria in your blood. That doesn't tell you why you're sick. Other people have the same bacteria in their blood and are not sick.

1–3 THE SEARCH REVIEWED

Man's view of the world around him in his first millions of years leaves trace of little change. We have to presume that prehistoric man, though he learned to make fire and spearheads, in general lived in superstitious fear of his natural surroundings.

Even in the recorded history of Sumeria, Babylonia, and Greece, where we find so much new art, architecture, and arithmetic, many natural happenings were attributed to the will of jealous gods and goddesses. Burning alive a nineteen-year-old virgin to satisfy the grain god was a long way from statistical inference.

About the middle of the second thousand years after Christ, people involved in scientific pursuits began to foster the somewhat radical idea that maybe all happenings might be explained on the basis of physical causes and effects. Indeed these men did begin to uncover what appeared to be the physical causes for a lot of what had previously been regarded as complete mysteries. Much of this confident view of things sprang from the at first frightening concept that the earth was not the center of the universe. Astronomers, with astonishing accuracy, began to predict the rising and setting times of the stars and the planets and the exact time when eclipses would take place. The earth revolved about the sun in accordance with the natural laws of inertia, motion, and gravity.

In the years that followed, a multitude of things were explained. It was possible, apparently, to say that the gravitational pull of the moon and sun caused the ocean tides, that passing a wire through a magnetic field caused a flow of electricity in the wire, that breathing fresh air into the lungs caused oxygen to be absorbed by the blood, that injecting into animals the toxins produced by diphtheria bacilli caused the animals to produce an antitoxin which, in turn, could be used to inoculate human beings against diphtheria.

1–4 STATISTICAL INFERENCE

Yet such explanations had a basic weakness: they were founded upon an idealization of the phenomena. The moon floating in frictionless space under the influence of the small number of bodies that compose the solar system is apparently an oversimplified situation when we face up to the gravitational forces exerted on the moon by the uncountable members of the universe. The diphtheria bacillus producing its toxins in a laboratory flask was a much simpler situation than the bacillus amidst the teeming microscopic world of organisms in the human blood. The velocity of a falling ball in a laboratory vacuum is much easier to predict than the movement of a nitrogen molecule in turbulent air. The daily changes in electrical voltage, the fluctuation of stock prices, the variations in successively machined bolt diameters, the monthly changes in sales volumes—even the subtly varied reactions of a human mind to a given stimulus—all supply information which allows us to come up with a probability statement only.

Man has finally made the bothersome discovery that happenings are affected by so many variables that it is probably impossible to isolate any one of them as an independent cause. Even the observer himself unavoidably affects the results of the experiment he is observing. It may even be, as some respected theoretical physicists say, that some things happen without any cause at all—electrons jumping from one orbit to another. Whatever the case, a scientific analysis of these situations may possibly be best accomplished by utilizing the concepts and techniques of statistical inference.

The coin comes heads up. The gambler draws an ace. Lightning strikes your home. These events may have causes, but the causes appear so complex that we are at a loss to fully explain why the events happened. It's chance, or should we call it luck? Anyway, do we have to take a fatalistic approach, shrugging our shoulders and saying, "That's the way the ball bounces"—as if no explanation left no grounds for prediction?

No! This is not so. Even though we don't know why, we do know what happened. And if we know what happened, then in effect we've got a sample. From sample events we can estimate, test, and correlate. Even a relatively small sample can afford us some confidence in our conclusions.

Maybe you are like Mitya in Dostoevski's *The Brothers Karamazov*, ". . . one of those who don't want millions but an answer to their questions." But the "why" questions are much too difficult. You can usually answer the "what" questions—what's in the sample—and statistical inference will help you to estimate what's in the whole. This may not satisfy Mitya, but it can be a lot of help to practical people in an uncertain world.

From a shuffled deck of ordinary playing cards, our chance of drawing the ace of hearts is $\frac{1}{52}$ or approximately 0.019. Now this probability concept is a lot better than shrugging off the result as simply the way the ball bounces. In truth, you can't tell what card will be drawn; you can't tell why the one drawn was drawn; but you can tell how probable it was that it would be drawn and how probable it is that it will be drawn again.

Maybe you're going to say, "It's a lot different with cards. You know what's in the pack." Good for you! That's just the reason why cards, dice, roulette wheels, and balls in a box are so often used to explain probability —because we know exactly what population we are drawing from. **Population** is a technical word in statistical inference and refers to the whole business from which we are getting our sample: the 52 different cards, the 6 faces of a die, the 38 compartments on a roulette wheel, the number of different colored balls in the box—these are the populations from which we draw. Though in life we usually don't know the populations, certain statistical concepts enable us to estimate them.

Now, things we can't explain are happening in our own lives all around us every day. Whether you take the super-complex cause theory or the no-cause theory, I don't care for the present. We have miles of decisions to make. My question clearly is: How can we take the situation we are interested in and analyze the details of it in such a way as to allow us to estimate a broader picture, to predict a probable future, and to decide what to do about it? Such is the goal of statistical inference.

1-5 CONCEPTS AND TECHNIQUES

The word *concepts* appears clearly in the title of this book. We are primarily concerned with the concepts behind confidence intervals, hypothesis testing, and correlation. In turn, these broad concepts of statistical inference are supported by more basic concepts, such as the binomial distribution, the normal distribution, chi-square, the analysis of variance.

In order to work out these basic concepts in specific cases, we have to make use of the appropriate statistical techniques and get into numerical calculations. But still this is not a skill course. Though we are going to use some calculatory skills for all they are worth, we shall use them not as ultimate objectives but as aids to improving our conceptual appreciation. This we have clearly in mind.

You will do a substantial amount of arithmetic, though you need very little algebra and no calculus. You will have to think about the exercises and decide what questions to ask. I don't want you to know how to do the exercises because you know what chapter they're attached to. That's not enough for a conceptual course. You should be understanding the concepts and accumulating techniques as you go along. You should be ready any time for questions on anything you've met before. In appropriate exercises for this conceptual objective you must decide what kind of a problem you've got, whether you have the necessary information to solve it, what technique you had better employ, and, finally, whether your decision is a reasonable one. That, as you will see, is a demanding assignment, but *decision* is really the name of our game.

Decision theory, springing from the philosophical implications of statistical inference, has become an area of very active study and speculation. People in business management, industry, government, scientific research, education, medicine, military strategy, have worked together on this new approach. Many books have dealt at length with the theory of decision making. The miraculous advantages of speedy and almost instantaneous processing of information by computers open up a new world of possibilities. But the basic concepts are still statistical.

We should admit at the start that in our efforts to come up with results on which to base our decisions in this very uncertain world, we have had to work with approximations. This is the way it goes. It may lack the traditional classroom security, but it is an unavoidable reality. We shall look closely at the nature of these approximations.

And I hope, as we go on, you will get real inspiration from seeing what man has done with the concepts and techniques of statistical inference—amidst all the variables, complexities, and uncertainties of his environment, in spite of his lack of causal explanations, in spite of the approximate nature of his computational results.

Estimation

Consider a huge suspension bridge like the Verrazano Narrows, Golden Gate, Mackinac, or George Washington Bridge. Its awesome beauty cannot fail to impress you. Then you may begin to wonder: "Can it really be safe—the whole massive construction hanging on wire strings?"

The George Washington Bridge crossing the Hudson River from New York City to Fort Lee, New Jersey, is completely supported by four steel cables, two along each side of the bridge. Each cable is 36 inches in diameter and is made up of 26,464 wires that are each 0.196 inches in diameter. From a concrete anchorage in New Jersey these great cables climb up over a 600-foot tower, sweep gracefully across the river to climb another 600-foot tower and reach back down to a concrete anchorage in Manhattan bedrock. Eight lanes of motor vehicle traffic and two footways plus an added tier of six more motor lanes are supported by these four cables.

If you consider all of the different wires in these cables, you can be very sure there will be some variation in their tensile strengths. But statistical theory tells us that the variability of the mean wire strength anywhere along the cable will be 1 over the square root of the number of wires times the variability of the strengths of single wires. The variability of the mean wire strength is only $1/\sqrt{26{,}464} = 1/163 = 0.006$ times the variability of the individual wires. That's why the bridge sweeps across the water so confidently.

This statistical estimation theory should impress you. It not only holds up suspension bridges, it is a relationship that makes sample means more dependable than the individual members of the sample. The mean weight of aspirin tablets or the mean length of 8-penny nails will show less variability than the individual tablets or nails. This is the basis for much of the work that comes later in this course.

Try to remember!

Hypothesis Testing

In the early 1950s the mysterious and terrifying poliomyelitis struck again. It struck hardest at the 5-to-9-year age group, accounting for 6% of their deaths. Many of those it did not kill it left as helpless cripples. The epidemic swept the country like a plague of old. Schools and playgrounds were closed. Man again lived in fear.

Then came the Salk vaccine. But how could the effectiveness of this unknown preventive be tested? How could the hypothesis that it was not effective be rejected? Such a life-or-death test must indeed be carried out on a very large scale. As the incidence of polio was about 50 per 100,000, a group of 40,000 children would only turn up about 20 on whom the test could reveal anything about the effectiveness of the vaccine. Much larger samples would be needed for a realistic reliability. Actually, at a cost of over 5 million dollars more than 1 million children were involved. It was the largest and most expensive testing of an hypothesis in medical history.

The test recommended by the National Foundation for Infantile Paralysis was carried out through a wide and elaborate employment of the basic concepts and techniques of statistical inference. The story is told in a very readable book by Paul Meier, *Statistics: A Guide to the Unknown.**

Try your library.

Correlation

Counseling services may examine your grade point average (GPA) in high school and predict your success in college.

(a) What have you got to say about the soundness of this procedure?
(b) Suggest a method for checking its effectiveness.
(c) Suppose your two GPAs are 3.4 and 2.9, respectively. Does this reveal any measure of their relationship?
(d) An answer to this could be: multiply 3.4 by 0.85 and add 0.01. Try it and you'll get 2.90. But it is only one of many possible routes to the same conclusion. Come up with another.
(e) What's your objection to applying such a rule in a specific case?
(f) You will probably admit that this is not a causal relationship. Cite some factors that might in specific cases make your predictors unreliable.
(g) Suggest a numerical way of expressing confidence in your estimators.

Statistical concepts and techniques will help a lot in these matters.

*San Francisco: Holden-Day, Inc., 1972.

· 2 ·

Sampling

By a small sample we may judge of the whole piece.

CERVANTES
Don Quixote (1615)

Firstly, a small sample in isolation tells us next to nothing about the quality of the batch from which it is drawn . . . . As any card player knows, it is extremely common for a hand of 5 cards to contain no face cards.

M. J. MORONEY
Facts from Figures (1956)

OBJECTIVES

To see the practical necessity for sampling and to appreciate some of the difficulties in getting a sample that is representative.

2–1 EXERCISES

1. Why do you suppose the whole from which the sample is taken is called the *population*? (What were government officials doing that forced Joseph and Mary to leave Nazareth for Bethlehem?)

2. How about the word *sample*? Try the dictionary.

3. Cite a situation where examining the whole population would be impractical and one where it would be impossible.

4. A sample is said to be *reliable* if other samples will produce similar results. How does this differ from being *representative*?

5. What is a *biased* sample? Give an example and explain why it is biased.

6. An inspector on a tack-stamping operation checks every 30 minutes the last 50 tacks produced by each machine. Why might this sampling technique produce biased samples?

7. Would a random sample of incomes in a city be the best way to estimate the average income? Would it be better to stratify the sampling to get a number of random samples from the upper-, middle-, and lower-income brackets proportional to the number of people in these brackets?

8. If an inspector checks every twist drill in a lot of 12,000 for burns, diameters, and lengths, why would you not expect him to catch every defective drill?

9. How could a careful study of 100 of these twist drills be more revealing than an examination of the whole lot?

10. What does the expression *random sample* mean to you?

11. You want to find out whether students in your college approve of the government's ABM (antiballistic missile) program.
    (a) State your objections to a balloting area.
    (b) How about mailing a ballot to each student?
    (c) Let's not be so negative; what would be a good way to do it?

12. To be useful in statistical inference what particular characteristics must a sample have? The sample must have those characteristics (**statistics**) which suggest corresponding characteristics (**parameters**) of the population.

13. Chickens have an enlarged pouch in the esophagus, called the crop, where food is stored and moistened, a preliminary digestive process. A professor of animal husbandry wanted to make a general statement about the contents of this crop for chickens in outdoor runs. The college had 25 outdoor runs with approximately 100 birds in each. How should the professor select a random sample of 50 birds? Random numbers?

14. Why can't you really draw without replacement a random sample from a small population? This seriously involves your own definition of random.

15. Why is it that you can keep drawing *random* samples from an infinite population?

16. Is the number of grains of sand on the beaches of the world infinite? How about the number of stars in the universe? Is infinity really a description not of the world of things but of abstractions? How about $\frac{1}{3} = 0.3333\ldots, \sqrt{2} = 1.414\ldots, \pi = 3.14159\ldots$?

17. Suppose you came from Mars and didn't know the color of the cards in a deck of ordinary playing cards. Taking a sample of 3, you get 2 diamonds and a heart; so you conclude that the cards are red. What *is* the probability that all 5 cards will be red?

18. Somewhere I read that in the United States a person is born every $8\frac{1}{2}$ seconds, a person dies every $16\frac{1}{2}$ seconds, a person immigrates every 71 seconds, and 1 emigrates every 23 minutes. Do you think this information came from sampling? Why?

19. Why must the conclusions from many *scientific* experiments depend on samples and statistical inferences from them?

20. There is definitely a correlation between scientific progress in certain areas and the applicability of statistical inference to those areas. Explain why statistical inference is difficult to apply in some areas of human study—sociology, education, psychology.

21. Your philosophy of life is a result of samplings, not usually a result of examining whole populations. Would you make this philosophy more dependable by examining small samples more closely or by loosely examining many more samples?

2–2 POPULATIONS AND REPRESENTATIVE SAMPLES

Population as used in statistical inference doesn't have to be people or animals. An inspector may take samples from populations made up of nuts or bolts or electronic devices. Other inspectors may take samples from populations of canned food, raw meat, or drugs. An educator may examine a sample of SAT scores (Scholastic Aptitude Tests) from the whole population of SATs taken in California in one year. A zoologist may study a sample of six chinchilla rabbits from a population of 10,000 in the Sunset Rabbitry. A population can be buttons, peas, or fleas.

Numerical descriptions of a population, like proportions, averages, or standard deviations, are called **parameters**.

It is often extravagant, if not impossible, to work with all the members of the population. For instance, if you test by firing them (called destructive sampling) every cartridge of 22-caliber made by an ammunition manufacturer, there won't be any left to sell. If a food processor weighs the drained contents of every can of tomatoes to be sure he is meeting the claimed weight on the label, the weighing may cost more than the tomatoes. If you check with every TV viewer in the United States to determine the average number of hours per week that he watches—but this is impractical! You'd *have* to sample.

Sample comes from the word *example*. And certainly as far as we are concerned, our sample must be an example of what the whole is like. The numerical descriptions of a sample (like proportions, averages, or standard deviations) are called **statistics** and it is these sample statistics that correspond to and suggest population parameters.

Sampling by looking at just one item from a population would not reveal very much. It is not likely that finding out how one person is going to vote will allow predicting how the whole election is going to come out. But a sizable sample could be more revealing. Knowing how 500 people out of 10,000 are going to vote obviously should produce something better to go on.

But it is not so obvious that, beyond a certain sample size, you reach a point of diminishing returns on your investment. For instance, testing the residual drug in the urine from 60 guinea pigs may be as revealing as testing it from 1000 guinea pigs. And this is a very important consideration where time, money, and human energy are concerned. A skillfully drawn small sample can be very revealing.

On the other hand, a poorly drawn large sample may be nonrepresentative and wholly misleading. A classic example of a nonrepresentative sample was that taken by a weekly magazine, *The Literary Digest*, previous to the 1936 Presidential election in which Landon was opposing Roosevelt. It was one of several polls but by far the largest. Ballots were mailed to 10 million voters, and over 2 million responded! The poll showed a victory for Landon. When election day came, Roosevelt swept every state except Maine and Vermont! (After the election Roosevelt used to refer to the respective senators as the ambassador from Vermont or the ambassador from Maine.)

Plainly the *Literary Digest* sample had not been representative of the whole population. The use of samples to suggest whole populations obviously requires that they must do just that. *The Digest* had randomly selected names from telephone listings and automobile registrations. Roosevelt, however, got many votes from people who didn't have telephones or automobiles at all. This may not be a complete explanation, but it does illustrate an unintentionally *biased* sample. The grand blunder hastened the collapse of *The Literary Digest*.

A fabric or a piece of wood is cut on the bias when the cut is made at an acute angle to the weave or grain. That is, it is cut on a slant. A biased sample is slanted (inclined in one direction, prejudiced, not representative). The *Digest* sample was biased in the direction of those people who had money enough to own cars or have telephones.

There is, admittedly, a risk in sampling. No sample can ever show everything that is in the population. But there is also a risk in trying to examine a whole population too. You don't know even then, because the results are too subject to human errors—and there are many opportunities to make human errors.

Certain industries, until the birth of quality control, used to check every article produced; this complete checking was called 100% inspection, and it was very expensive. Rows of inspectors sat at benches looking at every piece that had been produced. It was a deadly repetitive job. The inspectors were bored; their minds wandered; they grew weary. They failed to pick out all the defective pieces, not only late in the day, on bad days, or when something interesting from the office walked by, but even when the inspectors were at their best. Modern statistical quality control has emphasized that often the percent of defective pieces in large lots can be more accurately determined by sampling than by 100% inspection. More reliable conclusions can be reached by a close examination of samples than by a tedious, exhaustive survey of whole populations.

In many cases it is obviously impossible to examine whole populations —like the number of people in cars passing a given point on route U.S. 1, or the clarity of stamp cancellations at Tacoma, Washington post offices, or the blood count of Indianapolis citizens. When positive statements are made about the ratio of blonds to redheads in New York City, or the proportion of malarial carriers in Venezuela, or the proportion of mercury salts in swordfish caught off the New England coast—when such statements are made you *know* they must be estimations from samples.

You may prefer the relative security offered by games of chance where the populations are known: heads or tails, 6 faces of a die, 52 cards, 37 or 38 stops on a roulette wheel. But hold on a minute! Take a look at the buildings on the skyline of a big city. Any insurance companies represented? Certainly. But insurance companies have to work with populations much less well known than the population we draw from in ordinary games of chance. They often have to guess at the parameters of huge populations from some relatively small samples. But somebody must really know what he is doing, for the guessing is good and very profitable, as the skyline impressively states.

The technique of getting representative samples is a complete study in itself. Difficulties are encountered in every area of investigation. For instance, an industrial inspector wants to check roofing nails just made on a new stamping machine which has been running two hours. Can he get away with checking the last 50 just made? Of course not; if the machine had run erratically at times, there is no assurance whatever that the last 50 pieces would constitute a representative sample. If he takes samples at regular time periods, he may miss altogether any periodic malfunctioning of the machine in the intervals between his stops. Also, if the machine operator knows when the inspector is going to appear, he can adjust to make things look good at that particular time. Even changing his times and floor patterns may not avoid a predictable personal bias on the part of the inspector. But the inspector can attempt to avoid these biases by leaving his stops and samplings entirely up to chance.

2-3 RANDOM SAMPLES

The Selective Service System in the 1960s was subjected to much criticism of its lottery procedure of drawing names on slips from a revolvable drum. After the 1970 run the criticism was so severe that the National Bureau of Standards' Statistical Engineering Laboratory was called upon to help in the randomization procedure. The lab in turn came up with the idea of *two* drums, one containing birth dates and the other containing numbers which were to show the order of calling which accompanied the birth dates.

The following paragraph is from an article in *Science*, January 1971:

> The randomization procedures used for the 1971 draft lottery had two basic components. The first consisted of a multiple stage randomization which made use of a table of random permutations. The second consisted of a physical mixing, in public view, to give the lottery face validity and to appeal to the public's sense of what is random. These two components

are reminiscent of a commonly held lay interpretation of both American and English jurisprudence—that a court trial must not only be fair and just but must also give every appearance of being fair and just.*

If you *really* want to avoid sample biases, it isn't the appearance of randomization that counts. What counts is making sure that the sample really is random, that *at any time any item in the population has as much chance of being chosen as any other item in the population.*

One good way (usually difficult to accomplish) is to number all the items in the population and then use a table like Table I (Random Numbers) in our appendix. If your numbers are only one-digit numbers, you may go across any row or down any column, taking the numbers as they appear. If you have two or more digits, a somewhat similar procedure ought to be obvious. These random numbers follow no conceivable sequence; they are randomly determined.

You could play many games of chance with these numbers; start down the first column, taking 1, 2, 3, 4, 5, or 6 over and over again in the order in which they appear, and this will simulate rolling a die; take from two adjacent columns the numbers 00, 01, 02, . . . , 36, 37, 38 in the order they appear, and the results may simulate the results of spinning a roulette wheel.

Treated the other way around, dice and roulette wheels may be used in fashioning random number sequences. It is usually much more convenient to use a table. When the numbers have three digits or more, the random number table is indispensable. Yes, a table could have been used just as effectively in the Selective Service selection, but it would have been difficult and would not have been as convincing *visually* as the rotating drum.

Our treatment of estimation, hypothesis testing, and correlation is based upon random samples. Sampling techniques are of vital importance. We slight them here only because our book is not designed to be a technician's aid but to serve as enlightenment for the general student. So just don't forget that all our statistical computations are worse than useless if our samples do not to a fair degree represent the populations from which they came.

Beyond pure randomization, there are a number of special sampling refinements: systematic sampling, cluster sampling, stratified sampling, multiple sampling, sequential sampling. These methods you should know before you set about a research project. In fact, you'd better go to some library before you complete this course and stand in front of the "statistics" shelf; some day you may need its assistance.

*Stephen E. Fienberg, "Randomization and Social Affairs: The 1970 Draft Lottery," *Science*, Vol. 171, Jan. 22, 1971, p. 260, copyright © 1971 by the American Association for the Advancement of Science.

I and probably many students, professors, engineers, research scientists, and educators, both amateur and professional, have made extensive, but misleading, calculations based on samples that weren't truly random and representative because better sampling techniques were not made use of.

2–4 THE IMPORTANCE OF POPULATION SIZE

Now the random sample is not guaranteed to contain the information you need. But generally the larger you make the sample, the better the chance is that it will.

Small populations can give you trouble here. After drawing the first item without replacement, the remaining population can be substantially different for the second draw. If you wanted to date Mary, Alice, Ruth, or Muriel, and then Muriel was taken over by somebody else, the population you would have to choose from would be Mary, Alice, and Ruth—a bit more restricted. On the other hand, if you were to date from a large population, say from all the unmarried women over 18 in New York City, it wouldn't make much difference *statistically* if one woman named Muriel was removed from the population of eligibles. New York is a large population to date from.

Populations can even be infinite (without end, as compared with *finite*, which means with end, *finis*). If you want to go about picking whole numbers at random, what is the population you have to draw from? It is the infinite population of all possible numbers. Theoretically you could go on forever, selecting different numbers without affecting the size of the population from which you are drawing. (Is *forever* infinite?) How many times can you roll a pair of dice? Theoretically you can roll a pair of dice an infinite number of times—which would require living forever. No matter how many times you roll a pair of dice, the size of the population you are drawing from is not affected. If you want to take a sample of the grains of sand in the Sahara Desert, the population you draw from would be practically infinite, and even the removal of a carload of grains would not appreciably affect the total number of grains in the population.

I have brought this up because you ought to know, if you don't already, that sometimes it is important to consider the size of the population and sometimes it is not. If, to be random, every item must have the same likelihood of being drawn, then the population size has to be large enough to be very little affected by drawing items for the sample. For instance, we might insist that the sample be not more than 5% of the population. More of this later.

2–5 SUMMARY

No doubt you see the importance of sampling in the statistics of science, engineering, business, sociology, education, medicine—all areas of science and technology. The proportion of red cells in the blood, the average mineral contents in an ore deposit, the proportion of stored grain that is molding, the study of phone line loads, traffic flow, queuing in the retail business, average bank deposits in a given day, the proportion of consumers who prefer a certain hair wash, or the dangerous side effects accompanying the use of a drug to aid in the cure of cancer—all these must depend on sampling.

Great statistical contributions have been made in the physical sciences, the biological sciences, medicine. In sociology, political science, and economics, where perhaps the concern is even more critical, there has not been so much progress. Who knows; maybe you will develop a professional interest in statistics and produce something more effective in these areas.

Sampling and statistical inference lie at the core of the scientific approach. Unprofessionally, too, we are always sampling and inferring, trying to find out what life is like, trying to guess at its parameters. But we never know completely; only God is omniscient.

• 3 •

Chance—Probability

. . . Your Beethoven lives very unhappily, in constant conflict with nature and his Creator; oftentimes I have cursed the latter for making his creatures the sport of the most terrible chance . . . .

LUDWIG VAN BEETHOVEN
Letter to Carl Amenda (1801)

I shall never believe that God plays dice with the world.

ALBERT EINSTEIN (1947)

OBJECTIVES

To find a numerical way of indicating probabilities and a method for determining this probability in both theory and practice.

3–1 EXERCISES

1. What number scale is conventionally used to indicate degree of probability? Let's use three decimal places in the probabilities that follow.

2. Where do certainty and impossibility fit on this scale?

3. What do you mean when you say that something will *probably* happen? Give an example. Try giving this probability a numerical rating.

4. In rolling a die (one of a pair of dice), what is the probability of getting a 2? Remember: three decimal places.

5. What is the probability of getting a head with one flip of a coin? Not a head?

6. What is the probability of drawing the ace of spades from a full pack of ordinary playing cards? Of not getting it?

7. What is the probability, in flipping a coin, of getting a head or a tail? How about neither?

8. What is the probability, in rolling a die, of getting a 2 or a 3? Add separate independent probabilities for the probability of one or the other.

9. What is the probability of drawing the ace of hearts or the king of hearts? Add again.

10. This has been theoretical probability so far, based on our knowledge of the population. But more often we don't know the population we are drawing from and have to judge probabilities by samples. These probabilities are said to be determined empirically, as in auto accident rates, defective dishwashers per thousand, or mortality tables. Specifically, how would you find the probability that a randomly selected undergraduate student would go on to get his master's degree?

11. How could you estimate the proportion of cans of Top Choice clam chowder whose contents will be underweight ("net wt. 298 grams")?

12. Suppose your method finds the contents of 2 cans in 500 to be underweight. If you then randomly select one can of Top Choice clam chowder, what would you use as the probability that it will be underweight? Notice that for an individual occurrence, we use a probability that is

the same numerically as the theoretical population proportion or the same as the empirically arrived at relative frequency of its occurrence in samples. Think this over.

13. If you have the number of fair days in the last 10 years and divide it by 10 times 365 days, what can you say for this proportion (relative frequency) as a basis for predicting a fair day next Thursday?

14. Guess at the numerical probability that a tossed half dollar will stand on edge. Empirically, how could you check this?

15. What is the probability that today you will have a fatal car accident? Why is this, statistically, a poor question?

16. How large a sample do you have to take to *prove* an empirical probability?

17. What is your interpretation of the expression "a calculated risk"?

18. Give a few examples of numerical probability in use in areas other than in games of chance.

19. What is a binomial probability situation?

20. Carefully state what you think a random sample is.

3–2 A PROBABILITY SCALE

A summer thunderstorm, a pain in the back, a raise in pay, meeting the person you want to live with—can you give any of these a numerical probability? But living to be over seventy, having your house burn down, catching a trout weighing over one pound in Lake Winnipesaukee, New Hampshire—for these we have information that will allow us to estimate numerical probabilities. The scientist or the engineer has to use empirical probability measurements for the density of traffic on telephone lines or highways, thermal noise in electrical circuits, Brownian movement of particles in a liquid or a gas, excavation with nuclear explosives. (**Empirical** means from experiment or experience, as opposed to **theoretical**.)

Here is a possible way of indicating likelihood or **probability**. And please notice that from here and throughout the text we consistently use the industrial engineering practice of putting a zero before all decimal fractions. It is easy to miss decimal points (often causing somebody a lot of time, trouble, and money); the zero before the decimal point draws attention to the fact that a decimal point is coming.

PROBABILITY SCALE

1.000	Certainty
0.900	
0.800	
0.700	
0.600	
0.500	Fifty-fifty chance
0.400	
0.300	
0.200	
0.100	
0.000	Impossibility

Always $0.000 \leq P_{(\text{any event})} \leq 1.000$; that is, the probability of any event is equal to or greater than 0.000 but less than or equal to 1.000. All numerical probabilities are somewhere from 0.000 to 1.000, inclusive.

P stands for probability. The words or symbols in parentheses tell us the probability of what. The decimal that follows is a numerical register of its likelihood. This decimal fraction is obtained either theoretically (from a theory regarding the proportion of these "what's" in the population) or empirically (from experience with the relative frequency of the occurrence of these "what's" in a sample or samples).

Here are some probabilities:

Flip a coin:

$$P_{(\text{a head})} = 0.500 \quad (\text{"fifty-fifty"})$$
$$P_{(\text{a head or a tail})} = 1.000 \quad (\text{certainty})$$
$$P_{(\text{neither a head nor a tail})} = 0.000 \quad (\text{impossibility})$$

Name some other events that are certain or impossible.

Draw a card:

$$P_{(\text{a heart})} = 0.250$$

Roll a die:

$$P_{(\text{a three})} = 0.16666\ldots = 0.167$$

The weatherman says chances of rain tomorrow are 7 in 10:

$$P_{(\text{rain tomorrow})} = 0.700$$

An electrician claims chances of a faulty circuit are 3 in 1000:

$$P_{(\text{faulty circuit})} = 0.003$$

A doctor says chance of survival is 1 in 100:

$$P_{(\text{survival})} = 0.010$$

3–3 THEORETICAL PROBABILITY

Games of chance are good practicing grounds for probability exercises because we do know exactly what populations we are drawing our samples from and, consequently, can readily rate our probabilities numerically. Theoretically a coin has to come up heads or tails, and we believe each result to be equally likely. The chances are even. The odds are one to one. In the long run, half the time we should get a head; half the time we should get a tail.

$$P_{(\text{head})} = 0.500$$
$$P_{(\text{not a head})} = 1.000 - 0.500 = 0.500$$
$$P_{(\text{either head or tail})} = 0.500 + 0.500 = 1.000$$

(which you knew, of course)

Theoretically:

$$P_{(\text{any one face on a die})} = \tfrac{1}{6} = 0.167$$
$$P_{(\text{any one compartment on a roulette wheel})} = \tfrac{1}{38} = 0.026$$
$$P_{(\text{an ace})} = \tfrac{4}{52} = 0.077$$

We not only know what populations we are drawing from, but we also know that these populations are infinite, for you could, or somebody could, go on flipping coins, rolling dice, drawing cards (if each card is replaced after each drawing), or spinning roulette wheels forever. You remember our discussion in the last chapter on the importance of having the population large in relation to the sample. The chance of tossing a head is always 0.500 whatever happened on the previous toss. We shall keep our populations so large that the probability of drawing any item is not conditional, not dependent on what has been drawn; the probability of any item in our random sample will not change perceptibly during the sampling.

3–4 EMPIRICAL PROBABILITY

O.K. We flip the coin once and get a head. Does this prove that heads are more likely? Of course not. Well, flip the coin twice. Result: two heads. Does this prove that heads are more probable than tails? Perhaps we'd better go on. Next a head, then a tail, then another tail, until after 100 tosses we have had 46 heads and 54 tails. What does this prove?

Too small a sample, you say? All right, flip the coin 1000 times. Result: 561 heads, 439 tails. Does this give you grounds for saying that heads are more likely than tails?

Every so often the Sunday newspaper carries an article about somebody who has proved that the odds are not even because he has flipped many more tails than heads in 500,000 tries, or maybe a million tries. But, and this is important: you can never *prove anything* empirically, no matter how long you try, because you are only showing what has happened so far. You can't *prove* that the sun will come up tomorrow.

3–5 VITAL APPLICATIONS

Now you probably think this coin tossing is trivial stuff. I agree; it is, and greatly overused, too. But it is an obvious model for a probability situation; this helps. Suppose children in Los Angeles are dying in a polio epidemic associated with a new combination of viruses. In one year, out of 1000 cases, 503 die. Is the individual probability of death (503/1000) = 0.503? This is a binomial situation, like coin tossing, which simply means that it has to come out one way or the other: heads or tails; die or don't die; sink or swim.

Next, some medical research man comes along with the claim that he's found a vaccine that has a good chance of curing a child afflicted with the disease. Among the cries of the dying and the wails of the bereaved we feel compelled to give any possibility a chance. So 10 afflicted children are inoculated. Result: 6 die, 4 recover. Is the cure worse than none, the probability of surviving after administering the vaccine apparently being 0.400 as opposed to 0.503?

Not a big enough sample, you say? We try it on 90 more. Out of the total 100, just 30 die. Does this show that the probability of surviving after being given the new vaccine is 0.700? Should we postpone mass administration of the vaccine and continue experimenting on small samples? We are in desperate straits, for the fatal disease is carrying away children, maybe unnecessarily, all the time we wait. We've got to know.

Man has been making important decisions for ages, using insight, intuition, and prayer. But the concepts of statistical inference can help too —furnishing a systematic kind of hypothesis testing that has wide application. An introduction to the forming and testing of hypotheses is reason enough for a course in the basic concepts of statistical inference.

The theory of coin flipping does serve as a model for analyzing the effectiveness of a vaccine in halting certain diseases. And if you think I am exaggerating, you have forgotten that refinements on this probability concept in hypothesis testing were used on more than a million young children in 1954 at a direct cost of over $5,000,000 in order to evaluate the effectiveness of the Salk vaccine.

Probability is the nerve fiber of statistical inference—in estimation, hypothesis testing, and correlation as used in medicine, genetics, life insurance, baseball, ballistics, military strategy, industry, agriculture, forestry, nursing—you name it. And don't forget the overriding importance of getting a representative sample if getting a sample is necessary.

3–6 SUMMARY

If, before making any estimates about the population, you want something more definite than "possibly" or "very likely" or "not so likely," then you are going to have to count or measure pretty carefully and use a technique of inference quite a bit more exacting than "common sense." For instance, if you want to say in a decision about the doctor's antipoliomyelitic vaccine, that you are willing to accept a 0.010 probability of being wrong, then coming up with such a "calculated risk" is a good deal more demanding than guessing. (We are drawing, for all our wordiness, closer to the technical aspects of statistical inference.)

This course will not enable you to predict on the basis of a woman's honesty her probability of success in marriage, because if you've got nothing to measure or count, then you can't apply statistical techniques. But you can count or measure deaths per thousand for an insurance company, defects per batch for industry, hemoglobin deficiencies in hemorrhage cases for a hospital, or pregnancy cases after the pill for married (or unmarried) couples. In these areas prediction is possible empirically on the basis of the relative frequency of occurrence.

In a binomial situation an item either does or does not have a certain attribute: a coin shows heads or not, a patient is cured or not, a sales device impresses people or not, a person is schizoid or not. Of course a person could be schizoid, autistic, or—but we are going to put off considering several attributes at once until later (Chapter 14, Chi-Square). We are concerned now and will be concerned in the next chapter only with whether things do or do not have a certain attribute. This may be called a binomial situation.

• 4 •

Binomial Probability Distributions

Statistical manipulation and computation are only tools to be used after *intuitive understanding has been established. If the array of computational paraphernalia is introduced first, then more likely than not it will inhibit or kill the development of probabilistic reasoning.*

JEROME S. BRUNER
The Process of Education (1963)

OBJECTIVES

To recognize binomial situations,
to examine their probability distributions,
and to make use of a table
of binomial probabilities.

SYMBOLS AND FORMULAS FIRST APPEARING IN THIS CHAPTER:

π (pī) the proportion of a population that has a specified characteristic; this is also the probability of the specified characteristic in a sample of one from that known population

n the number of tries; the sample size

x the number of items in the sample that show the specified characteristic

4–1 EXERCISES

1. Give an example of a binomial situation. What could be a multinomial situation? Is there such a thing as a monomial situation?

2. List the probabilities that the ball in our pinball model will reach each of the 8 pins in the seventh row. You'd better use the table of binomial probabilities.

3. Graph a probability distribution of this information.

4. This distribution is symmetrical. Explain carefully what that means.

5. If you toss 9 coins, what is the probability that 3 (not 3 or more, but exactly 3) will fall heads up? What about the same coin 9 times? Getting 3 is ambiguous; you can't tell whether it means at least 3 or exactly 3; that is why we said "exactly."

6. What is the probability of getting 3 or *more* heads with one flip of 9 coins?

7. What is the probability of getting more than 3 heads with 9 coins?

8. What is the probability of getting 3 or less with 9?

9. How about less than 3 with 9?

10. What about the probability of getting between 3 and 5 heads, inclusive, in 8 tosses of a coin?

11. Suppose you have a huge basket full of quarter-inch beads, 0.40 being red. What's the probability of getting exactly 4 red beads in a random sample of 10?

12. A hardware retailer finds that 5% of his paintbrushes from one manufacturing company have been defective (losing an excessive number of bristles); what is the probability that an interior decorator will find one (not one or more, but exactly one) defective brush in 15 of these brushes?

13. Draw a playing card and replace it. Shuffle the cards after each draw. What is the probability that you will get no hearts in 13 tries? As $\pi = (13/52) = 0.25$, you'll have to guess at a value, say halfway between the probabilities for $\pi = 0.20$ and $\pi = 0.30$. The value won't be just right

because when you go halfway between two π's, that does not get you exactly halfway between the probabilities.

14. If the individual effectiveness in relieving headaches of some advertised pill is 80% (a pretty difficult thing to measure, but not so difficult to rig up a convincing TV picture of), then how likely would it be that all 10 out of 10 selected users would claim relief?

15. Among tomato soup cans marked "Net weight 12 oz," it is claimed that 5 in 100 will be underweight. What is the probability of getting exactly 1 underweight can in a random sample of 10?

16. How could you go about finding the binomial probabilities for $n = 16$?

17. Pills for hypertension relief are claimed to contain 80% hydralazine hydrochloride. In a sample of 15 pills, how many will contain more than 80% hydralazine hydrochloride?

18. Would the continued occurrence of tails in tossing of coins make heads more probable on the next toss? What is "the law of averages"?

19. Make up a problem involving a binomial situation. Get an answer for it.

4-2 A BINOMIAL PROBABILITY DISTRIBUTION

At the top of the next page is a diagram of a special pinball machine, gently inclined so that the black ball rolls downward. Look at it thoughtfully. Don't hurry. Do you see that we have really set up a kind of probability distribution? The channel guides the black ball to the first pin. No uncertainty here. No distribution of probabilities.

$$P_{(\text{hitting first pin})} = 1.000$$

But next the ball goes either to the right or to the left; this is a binomial situation, and the first row appears as a probability distribution. Now, if the ball is perfectly spherical, the pin perfectly cylindrical, the ball material flawless, etc., then the ball will have the same probability of going either side of the pin.

$$P_{(\text{to the right})} = 0.500$$
$$P_{(\text{to the left})} = 0.500$$

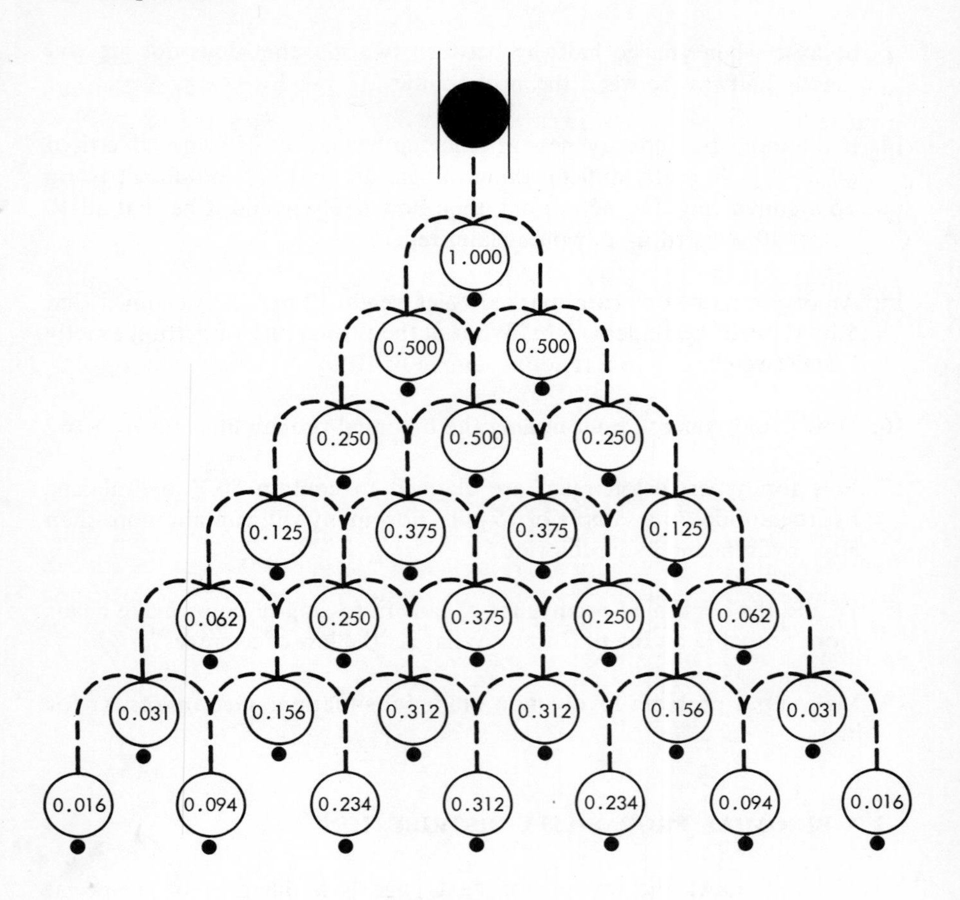

Pinball machine
(probabilities shown)

The probability of going one way or the other at any pin will be constant (0.500).

As you can see, one pin produces two paths to the two pins of the first row. Don't count the first pin as a row. Then two pins produce three paths to the three pins of the second row. Next, three pins produce four paths to four pins. And, in our diagram, on up to seven pins in the sixth row.

The probabilities recorded in the diagram could have been determined by simply tracing the possible paths. For instance, in the second row there are three pins. There are two ways from the first row to reach the middle pin of the second row, but only one way to reach each of the end pins, so the probability of reaching the middle pin is twice that of reaching the end pins. The total probabilities for each row must add up to 1.000 (the ball must reach one pin or another). In the second row, $\frac{1}{4} + \frac{2}{4} + \frac{1}{4}$ or $0.250 + 0.500 + 0.250 = 1$.

(Think about the total number of routes to any row as equal to 2 raised to a power that is the row number: second row, 2^2 or 4; fifth row, 2^5 or 32.) The probability of hitting any one pin then is the number of paths to that pin, divided by 2 to the power of the row number.

$$P_{(\text{hitting fifth pin in fifth row})} = \frac{5}{32} = 0.156$$

$$P_{(\text{hitting fifth pin in sixth row})} = \frac{15}{64} = 0.234$$

$$P_{(\text{hitting sixth pin in sixth row})} = \frac{6}{64} = 0.094$$

Now consider tossing coins. As with the ball (right or left), there are only two possible outcomes (heads or tails). We again have a binomial situation. Also, the probability of either outcome is again 0.500. Put the pinball diagram to work, using the row number (n) for the number of coins, using the first pin for all heads, the second pin for $n - 1$ heads, etc., down to the last pin for no heads.

$$P_{(\text{all heads, 6 coins})} = 0.016$$

$$P_{(\text{5 heads, 6 coins})} = 0.094$$

$$P_{(\text{3 heads, 5 coins})} = 0.312$$

$$P_{(\text{no heads, 2 coins})} = 0.250$$

This distribution will hold for any binomial situation where the individual probability is 0.500 each time. If the probability of your hitting bottom from a high diving board is 0.500 every dive, then the probability of doing it 5 times out of 6 dives is 0.094, though we'd expect you to learn a little something from experience, so that the individual probability of hitting bottom (0.500) would not stay constant. Balls and coins, of course, don't learn.

If something can happen only one of two ways and the probability each time is constant, the situation is a binomial one.

π(pī) = constant individual probability

n = number of tries

x = specified number of same results

Twice we have used

$$\pi = 0.500$$
$$n = 6$$
$$x = 5$$

(coins, dives) and got $P_{(5 \text{ out of } 6)} = 0.094$.

We could present the probability distribution for $\pi = 0.500$ and $n = 6$ in the following manner.

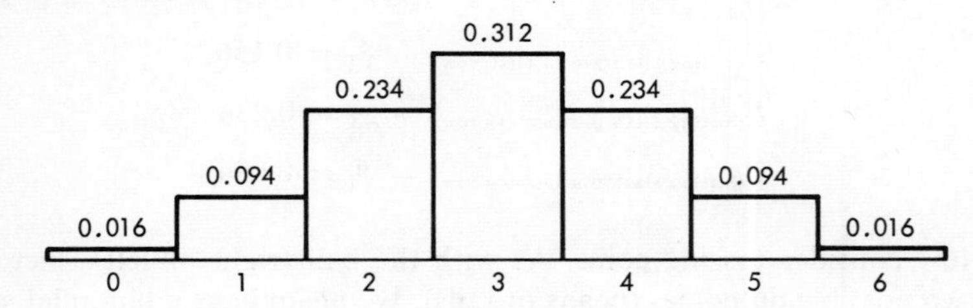

Probability distribution for different x's when $\pi = 0.500$ and n = 6

This is called a **histogram**. The heights of the rectangles are drawn proportional to the probabilities. This makes the areas proportional to the probabilities.

Here it is for $\pi = 0.500$ and $n = 4$.

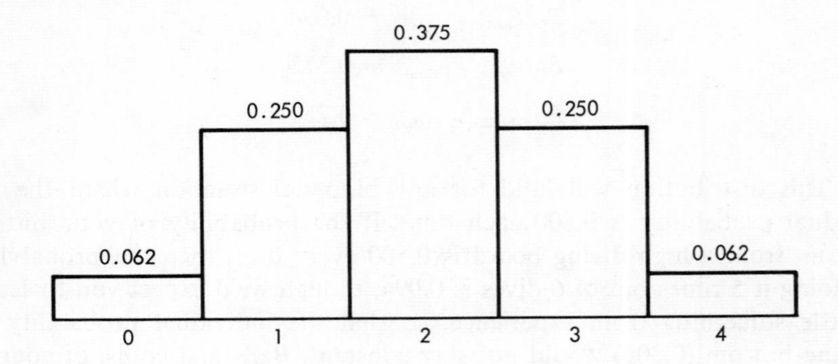

Probability distribution for different x's when $\pi = 0.500$ and n = 4

Notice that both our graphs are symmetrical.

4–3 THE BINOMIAL TABLES

Our distributions will work for any constant binomial probability with $\pi = 0.500$, whether it be pins, heads, hearts, or automobile fatalities. But our distributions go up to only $n = 6$.

What is the probability of 4 heads in 10 tries?

Go to Table II in our appendix (page 227). This table goes up to $n = 15$. Come down the π column headed 0.500 to the $n = 10$ block of figures. $x = 4$ shows 0.205.

$$P_{(4 \text{ heads, } 10 \text{ coins})} = 0.205$$

Go back to 5 out of 6. $\pi = 0.500$. Come down the 0.500 column to block $n = 6$ and then come down to row $x = 5$. There you have it again:

$$P_{(5 \text{ out of } 6)} = 0.094.$$

Just like that!

Now we've got the problem of what to do if π is not 0.500.

Dented Fenders

An authority states that the most likely car fender to be dented is the right-rear. Selecting cars with only one dented fender, he said 30% would have the right rear dented (next in likelihood, the right front).

Suppose you take a random check of 10 cars with one dented fender each. What is the probability that 4 of them will have the right-rear fender dented?

Go back to Table II in the appendix. Your table lists π for 0.050, 0.100, 0.200, 0.300, 0.400, 0.500, 0.600, 0.700, 0.800, 0.900, and 0.950. Come down the 0.300 column instead of, as we did before, the 0.500 column, until you get to the $n = 10$ block and go out the $x = 4$ row.

$$P_{(4 \text{ out of } 10, \text{ dented right rear})} = 0.200$$

Really each column in each block represents a probability distribution. In this case, our results are shown in the following diagram.

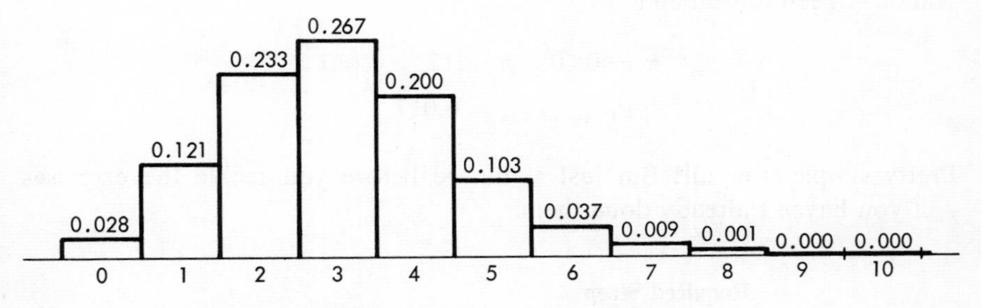

Binomial probability distribution for x's ($\pi = 0.30$, $n = 10$)

When π is not equal to 0.500, the distribution will not be symmetrical, as you can see.

Looking at our graph we can readily select probabilities:

$$P_{(6\text{ out of }10)} = 0.037$$

$$P_{(6\text{ or }7\text{ out of }10)} = 0.037 + 0.009 = 0.046$$

$$P_{(\text{more than }7\text{ out of }10)} = 0.001 + 0.000 + 0.000 = 0.001$$

$$P_{(\text{less than }3\text{ out of }10)} = 0.233 + 0.121 + 0.028 = 0.382$$

The proportion of what we are looking for in the whole population is π. The probability of getting what we are looking for on a single draw is, therefore, π. We can change from one π to another by selecting the appropriate column in our table of binomial probabilities.

Community College Students at State University

At a certain community college, 70% of the students who transfer to the state university show GPAs over 1.5 in the first semester at the university (not really a very revealing piece of information). In a random sample of 15 of these students, what is the probability that 13 or more will have GPAs over 1.5 in the first semester?

Referring again to Table II: block $n = 15$, column $\pi = 0.700$, rows $x = 13, 14, 15$.

$$P_{(13\text{ or more})} = 0.092 + 0.031 + 0.005 = 0.128$$

Unwanted Pregnancies in U.S.A.

A national survey reports that about 40% of the natural increase in United States population is accounted for by unwanted pregnancies. In a random sample of 12 childbirths, what is the probability one will have been unwanted?

$$\pi = 0.40, \quad n = 12, \quad x = 1$$

$$P_{(1\text{ out of }12)} = 0.017$$

Pretty simple after all! But just a minute before you tackle the exercises —if you haven't already done them.

Required Sleep

A health specialist tells Jones that he must get at least 8 hours sleep every night. Eight hours is $\frac{1}{3}$ of 24, or 0.333. The specialist checks Jones on 12 randomly selected nights. What is the probability Jones will have slept 8 or more hours on 5 of these nights?

Of course we can't do it! None of these figures shows the individual probability of Jones sleeping 8 or more hours on a single night (π). You must have π (the probability in a single instance) in order to do this type of problem. Really think about it: The proportion *must* be the individual probability, not some other misleading proportion.

In attacking your problems, always first frame very specifically the question being asked. Then determine whether you have the information necessary to answer it. If you do, now is the time to go about your calculations, not ever before this.

4-4 SUMMARY

A situation is binomial when the individual happenings can occur in only one of two ways (heads or tails, black or white beans, soil radioactive or not) while their individual probabilities remain constant. From binomial probability distributions, given the individual probability (π) and the sample size (n), we can state the probability of either one of these two ways occurring any specified number of times (x). Binomial probabilities thus become a very useful basic concept.

So far, however, we have been limited to samples no larger than 15. Next, in Chapter 5, we are going to deal carefully with the famous *normal* distribution. In Chapter 6 we shall see how this normal distribution may very often be used as a model for a binomial distribution. This will enable us to handle samples of more than 15, greatly increasing the usefulness of our binomial probability concept.

If you are curious about the mathematical theory behind binomial probabilities, go to the binomial expansion in an algebra text. See how the exponents in the expansions serve as our x's. If you are even more curious, look up permutations and combinations and discover how combination theory is used to give you these x's.

• 5 •

Frequency Distributions—Normal Distributions

. . . We cannot get more out of the mathematical mill than we put into it, though we may get it in a form infinitely more useful for our purpose.

JOHN HOPKINSON (1894)

OBJECTIVES

To appreciate what a frequency distribution shows, to examine particularly the so-called normal frequency distribution, and to set up a standard normal probability distribution that can help us with probability estimates.

SYMBOLS AND FORMULAS FIRST USED IN THIS CHAPTER:

$\bar{x} = \frac{\Sigma x}{n}$	(x bar) the mean of the sample
$\mu = \frac{\Sigma x}{N}$	(mū) the mean of the population
$s = \sqrt{\frac{\Sigma(x - \bar{x})^2}{n}}$	the standard deviation of the sample
$\sigma = \sqrt{\frac{\Sigma(x - \mu)^2}{N}}$	(sigma) the standard deviation of the population
$z = \frac{x - \bar{x}}{s}$	the number of sample standard deviations of a single value (x) from the sample mean
$z = \frac{x - \mu}{\sigma}$	the number of population standard deviations of a single value (x) from the population mean

5–1 EXERCISES

1. A new antibiotic suspected of causing an undesirable rise in temperature is given to 100 patients suffering from intestinal infections.

Temperature-Interval Boundaries	Frequencies
98.0– 99.0	3
99.0–100.0	10
100.0–101.0	25
101.0–102.0	35
102.0–103.0	20
103.0–104.0	5
104.0–105.0	2
	100

Draw a frequency histogram for this information. Give it a complete title. Show clearly the frequencies and the interval limits (not the boundaries).

2. The standard normal distribution has $\mu = 0$ and $\sigma = 1$. For any z in our standard normal table, the probability of a random variable falling between μ and plus or minus that z is shown. Explain how you can still use the same table for a normal distribution of such units as inches, gallons, microvolts.

3. What is the probability that a variable randomly selected from a normal distribution will fall within
   (a) $\mu \pm 1.00\sigma$?
   (b) $\mu \pm 2.00\sigma$?
   (c) $\mu \pm 3.00\sigma$?

4. Find the probability that a random selection from a normal distribution will fall between
   (a) $z = 0$ and $z = 1.50$.
   (b) $z = 0$ and $z = 2.31$.

5. What is the probability that a random selection from a normal distribution will fall between
   (a) $z = 0$ and $z = -1.50$?
   (b) $z = 0$ and $z = -2.97$?

6. What part of the area under a normal curve lies between
   (a) $z = 1.5$ and $z = -1.5$?
   (b) $z = 2.70$ and $z = -1.78$?

7. Find the probability that a randomly selected variable from a normal distribution will fall
   (a) Beyond (to the right of) $z = 1.96$.
   (b) Beyond (to the left of) $z = -2.33$ or beyond (to the right of) $z = 2.33$.

8. What z for a normal distribution will give 0.3485 of the area between it and the mean? What $-z$ will give 0.4778 of the normal area between it and the mean?

9. What z to the right and also to the left of the mean of a normal distribution will locate two lines (obviously the same distance away from the mean) that will include 98% of the area?

10. The mean PSAT score for some high school juniors in a certain year was 35 with a standard deviation of 12.
    (a) What proportion got under 40, assuming a normal distribution?
    (b) Find an interval, with boundaries the same distance above and below the mean, that will include 0.50 of all scores.

11. Suppose the fatality rate for patients having pneumonia after being treated with sulfapyridine is 0.05. What is the probability that 3 out of 10 in a random sample will die?

12. A large group of male industrial workers 30 to 39 years of age was checked for systolic blood pressure, the mean showing 125 millimeters and the standard deviation being 15 millimeters. Assuming a normal distribution, find the $\pm 3\sigma$ interval. Notice that there is no mention of what time of day it was nor of what the workers had been doing. How can blood pressure be measured in linear units?

13. A portable radio was designed on the hypothesis that 70% of the purchasers would be women. If this is so, what is the probability that in a sample of 400 purchasers less than 270 will be women? Assume a normal distribution with $\sigma = 9.2$.

14. If the mean of a distribution of the life of light bulbs is 1000 hours with a standard deviation of 100 hours, what percent will last less than 900 hours? Assume a normal distribution.

15. If you roll a die two consecutive times, in what way are these results related?

16. The life expectancy of wooden telephone poles has a mean of 10.5 years with a standard deviation of 3.3 years. Show a life interval you think will include 95% of the poles, if the distribution is normal.

17. A manufacturer knows that the number of employees absent from work on an "average" day is approximately normally distributed with $\sigma = 5$. What is the probability that on a randomly selected day more than 30 employees will be absent?

18. A petroleum company averages 2 strikes for every 10 wildcat wells it drills. Seven wildcat wells are drilled in widely separated areas. What is the probability that 2 or more will be strikes?

5–2 FREQUENCY DISTRIBUTIONS

Between the first and second periods of a basketball game, a thief stole the money bag from the ticket office, ran across the basketball floor, and disappeared through a rear fire exit into the darkness of night.

The police arrived amidst screaming sirens. They wanted a description of the man who had made his escape in plain sight of so many spectators.

How tall was he? First answer: "Just over 6 feet." Second answer: "About $5\frac{1}{2}$ feet."

Inspector Columbus decided to ask eight more spectators to estimate the thief's height to the nearest half foot. Results of sample of 10: 6, $5\frac{1}{2}$, $5\frac{1}{2}$, $5\frac{1}{2}$, 6, 6, 5, 6, $5\frac{1}{2}$, 6.

Shown as a frequency distribution table:

Estimated Heights in Feet	**Frequencies**
6	5
$5\frac{1}{2}$	4
5	1
	$n = 10$

These same measurements are shown as a frequency distribution graph (histogram: story picture) at the top of the facing page. What measurement occurred most often? 6 feet. This is called the **mode**. (Fashion styles that occur most often are said to be the mode.) Is this the best estimate?

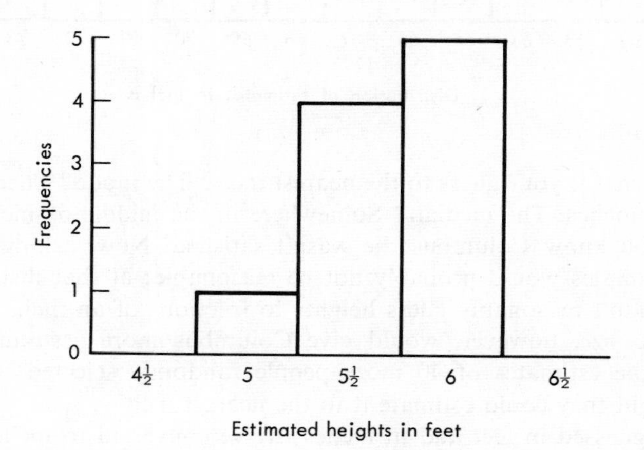

You could take a height which has half of the counts below it and half of the counts above it. That could be $5\frac{3}{4}$ feet. This middle number is called the **median**. (The median strip is the area between dual highways.) Is this a better estimate?

Inspector Columbus went back to the same 10 people and asked them to try estimating to the nearest inch. Results: 71, 66, 65, 68, 73, 74, 62, 70, 68, 70 inches.

Estimate to Nearest Inch	**Frequency**
74	1
73	1
72	0
71	1
70	2
69	0
68	2
67	0
66	1
65	1
64	0
63	0
62	1
	$n = 10$

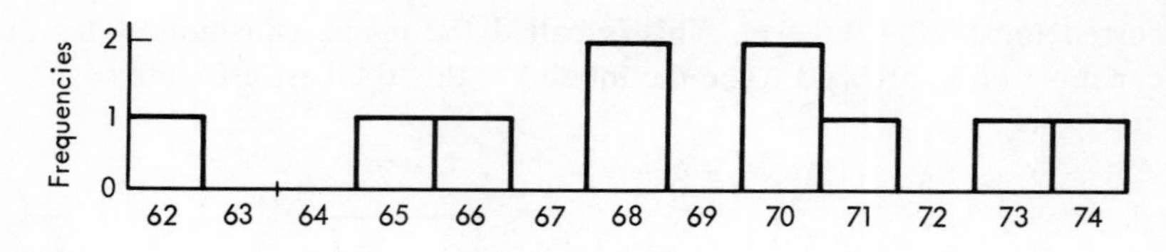

Now what is your guess to the nearest inch? The mode? There are two: 70 and 68 inches. The median? Somewhere in the middle 68-inch interval.

But you know Columbus: he wasn't satisfied. Now, asking for more *precise* estimates would probably not be reasonable; at that distance spectators couldn't reasonably guess heights to fractions of an inch. Increasing the sample size, however, would give Columbus more assurance. So he recorded the estimates of 40 more people randomly selected from those who thought they could estimate it to the nearest inch.

They guessed in feet and in inches; we've converted to inches.

First Sample
71
66
65
68
73
74
62
70
68
70

Additional 40

65	62	73	70
62	70	65	74
63	73	68	73
65	73	73	76
73	60	71	68
70	70	81	79
62	78	66	78
68	63	70	76
65	71	70	71
71	71	73	71

TALLY SHEET

81	/	1	
80			
79	/	1	
78	//	2	
77			
76	//	2	
75			
74	//	2	
73	~~////~~ ///	8	
72			
71	~~////~~ //	7	
70	~~////~~ ///	8	← mode, median
69			
68	~~////~~	5	
67			
66	//	2	
65	~~////~~	5	
64			
63	//	2	
62	////	4	
61			
60	/	1	
		$n = 50$	

Mode: 70 inches
Median: in the 70-inch interval

Here is another histogram showing the frequency distribution.

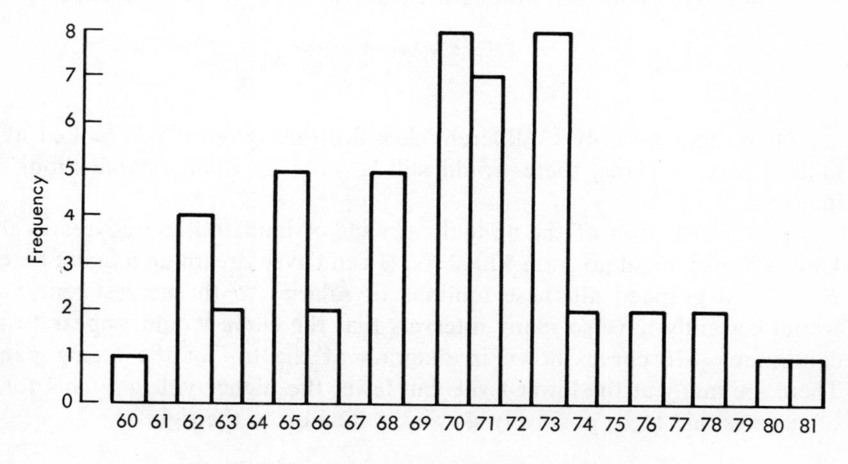

Distribution of 50 estimates of thief's height in inches

Inspector Columbus settled for 5′10″ (70 inches). We'll take a look at the *average* height a little later.

In regard to the number of intervals, the most we had was 22, when the spectators estimated to the nearest inch. Do you see that the more intervals we have in a given space, the narrower they must become? What is the limit? When the intervals become infinitesimally small and the sample is infinitely large, the steps will disappear, giving us a smooth curve—a **continuous** curve. Archimedes used this argument in the third century before Christ. He asked what would be the limit approached if one kept increasing the number of sides on a polygon—pentagon, hexagon, septagon, octagon (visualize it), hundred sides, thousand sides, million sides. The individual sides become so small that the ultimate theoretical limit becomes a smooth polygon or circle. So, likewise, in a frequency distribution as the number of intervals is increased, their width becomes smaller and smaller until the distribution shows no steps at all and looks like a smooth *continuous* curve.

Frequency distributions come in all shapes. Let's look at a few more.

Take first, the relatively simple population formed by a pack of ordinary playing cards. Consider the number of cards with the same face values. Of course! 4 of each. The distribution will be rectangular in shape.

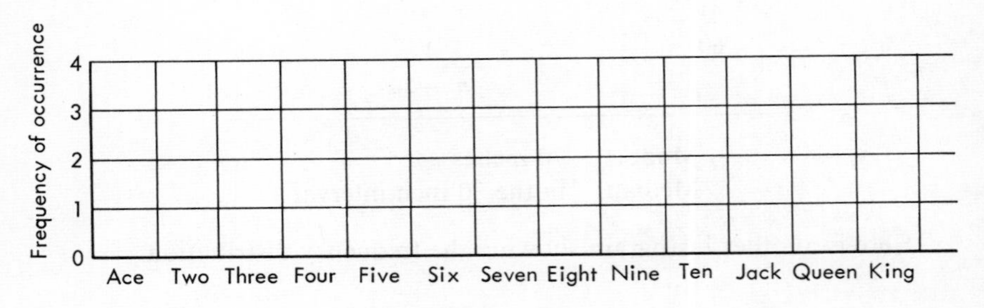

Here there are only 13 different classifications. Even if you looked at a million playing cards, there would still be only 13 value classifications or intervals.

The distribution of the annual earnings of industrial employees in the United States might go from $7000 to? (I can't even dream up a figure here.) *But* if you graphed all these millions of salaries to the nearest cent, you would certainly have so many intervals that the curve would appear to be continuous. See curve shown in diagram at the top of the facing page. There are many at the lower levels but fewer the higher you go. This form of curve is said to be positively skewed, tailing off to the right.

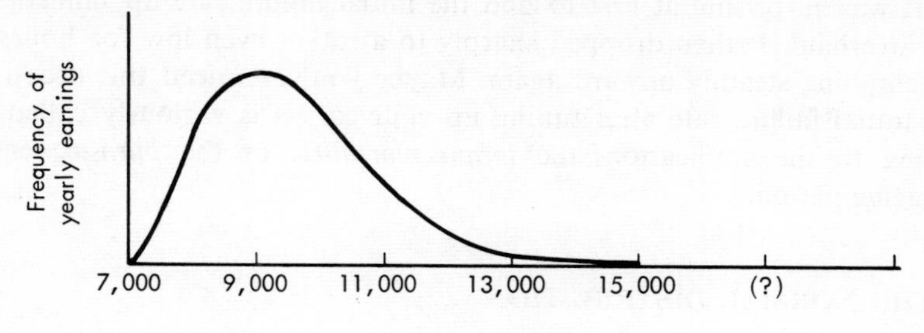

The age of people in the United States at death would, considering the population size, certainly appear to be continuous. This curve might be skewed to the left, with a slight secondary peak for deaths at birth.

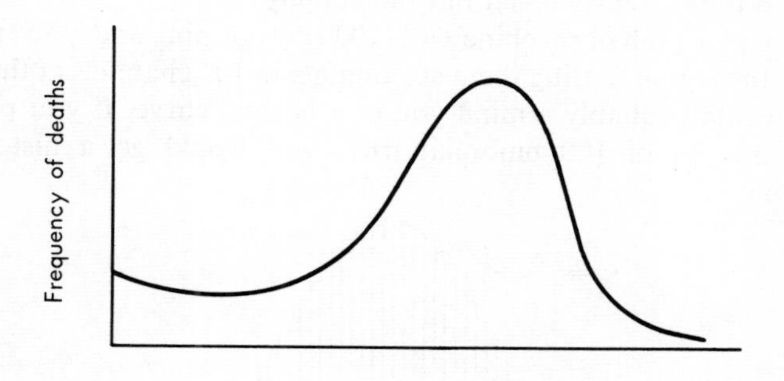

An early statistical study was done on the failure rate in military aircraft after overhaul during World War I by the U.S. Air Force, and it revealed the now well-known "bathtub" curve.

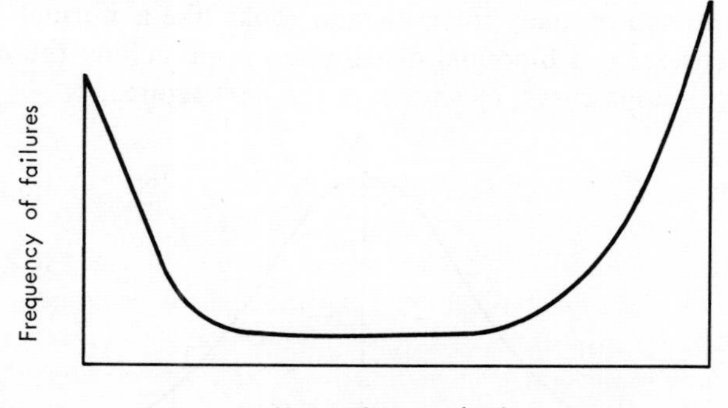

It was surprising at first to find the initial failure rate up immediately after overhaul. It then dropped sharply to a rather even low for hours before climbing steadily upward again. Maybe you've noticed this occasional high initial failure rate after tuning up your car. It is variously called, according to the application, the *infant mortality*, or the *burn-in*, or the *debugging* period.

5–3 THE NORMAL DISTRIBUTION

There are an unlimited number of distribution shapes. The one we are particularly interested in is called the *normal* frequency curve. I think it looks like Napoleon's hat, but it is generally described as bell-shaped. Most of the occurrences are at or near the center; then their frequencies fall off on both sides, at first rapidly and then more and more slowly, until the extremes reach out indefinitely.

If you had a pinball machine with 100 rows of pins and you ran a great many balls through it, letting them accumulate in 101 channels at the bottom, the result would probably remind you of a normal curve. If you plotted the *theoretical* results of 100 binomial tries, you would get a histogram, as shown below.

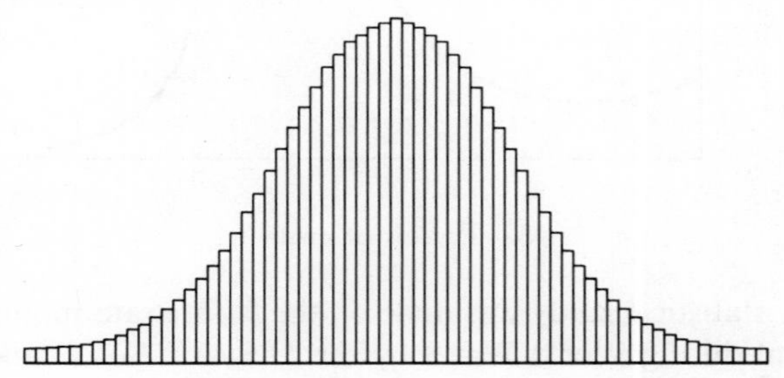

The graph shows many intervals and looks like a normal distribution. We could consider it a binomial distribution approaching the normal. The limit is a continuous curve, as shown in the next figure.

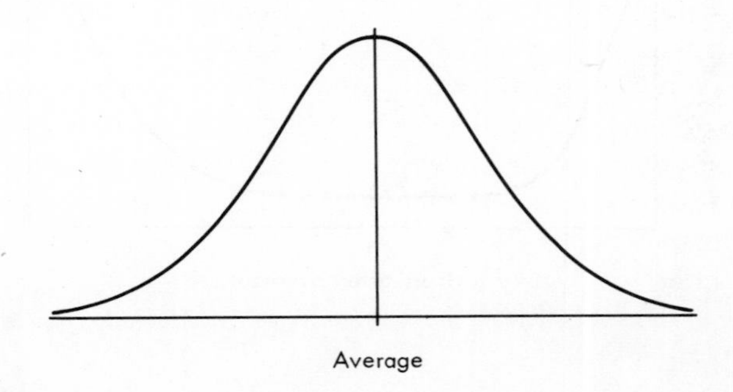

The normal curve, when first studied in the eighteenth century, was known as the *normal curve of errors*. It had been observed that repeated, precise measurements of the same thing showed a bell-shaped variation, piling up at the center but falling away indefinitely on both sides. The ends appear never to touch the horizontal axis. (This means that theoretically no extreme results are impossible.)

Your algebra certainly showed you a straight line for an equation in two variables like $y = 2x + 3$ and a curve for an equation like $y = x^2 + 2x + 3$. The normal curve can also be expressed as a relationship between two variables x and y:

$$y = \frac{1}{\sqrt{2\pi}} e^{\frac{-x^2}{2}}$$

Take a sightseer's view and pass along.

Likenesses to the bell-shaped curve keep turning up in practice. The weights of 1000 college women randomly selected, or the heights of 500 grass blades randomly selected, or the results that 100 randomly selected people get in measuring the length of the same table to the nearest $\frac{1}{10}$ of an inch will probably be somewhat bell-shaped when presented as frequency distributions.

Quoting from Sir Francis Galton (1822–1911), famous mathematician and cousin of Charles Darwin:

> I know of scarcely anything so apt to impress the imagination as the wonderful form of cosmic order expressed by the "Law of Frequency of Error." The law would have been personified by the Greeks and deified, if they had known it. It reigns with serenity and in complete self-effacement amidst the wildest confusion. The hunger of the mob and the greater the apparent anarchy, the more perfect its sway. It is the supreme law of Unreason. Whenever a large sample of chaotic elements are taken in hand and marshalled in the order of their magnitude, an unsuspected and most beautiful form of regularity proves to have been latent all along.*

But more important to us than the frequency distribution of large numbers of individual measurements is the truth that under the right circumstances, the normal curve does describe accurately enough the variation of the sample counts, proportions, and measurements used in statistical inference. This phenomenon can be supported mathematically. Under certain conditions the distribution of sample means from a given population has the shape of a normal curve. So does the distribution of sample proportions. These are surprisingly dependable statistical regularities in the midst of life's uncertainties.

*As quoted in Helen M. Walker, *Elementary Statistical Methods* (New York: Henry Holt and Company, 1943), p. 166.

5–4 STATISTICS AND PARAMETERS

It bears repeating that the mean and other characteristics of samples are called statistics, while the same characteristics of populations are called parameters.

The **mean** of certain measurements in a sample is the average of them. It is used to indicate the central tendency. It is probably the commonest statistic used in everyday life, though the **median** and **mode** are common statistics, too. To get the mean, divide the sum of all the measurements by the number of them. Labeling the different measurements $x_1, x_2, x_3, \ldots,$ x_n (n being the number of measurements) and calling the mean $\bar{x}$ (x bar), produces this formula for a sample mean:

$$\bar{x} = \frac{x_1 + x_2 + x_3 + \cdots + x_n}{n}$$

It is easier to use the Greek capital sigma, Σ (our S), to indicate summation (addition) and write only

$$\bar{x} = \frac{\Sigma x}{n}$$

In this case, x means any individual x.

We use these signs to help you, not to impress you with the mathematical wizardry of the author. Signs are the shorthand of mathematics, not a secret code.

If you are concerned with the mean as a parameter of the population instead of as a statistic from the sample, write the formula with the Greek small mu, μ (our m), for the population mean and N for the population size.

$$\mu = \frac{\Sigma x}{N}$$

This change in symbols enables you to indicate clearly whether you think you are dealing with a sample or a population, which is important, as you will see.

Don't just brush over this quickly in your haste to get back to the exercises. Stop and look at it. Do you know what each symbol means? Getting it clear now may prevent a lot of trouble later on. And I have even tried to simplify conventional formulas. A more mathematically acceptable formula in this case would be:

$$\bar{x} = \frac{\sum_{i=1}^{n} x_i}{n}$$

But for the purpose of this book, I prefer to take some liberties with conventional constructions.

There will be times when whether you consider the information to come from a population or a sample will depend upon how you look at it or how you agree to look at it. I come home with 10 pickling cucumbers in a bag. I take out a couple (the sample) to determine what the other 8 in the population look like. Or I look at the 10 (the sample) in an effort to estimate what the store supply (the population) looks like.

Computing the mean gives us the central tendency of a distribution, but that is not going to be enough. We want to know more than that about our frequency distribution. For instance, what is its spread or *range*? Estimates of the thief's height showed a range of $5\frac{1}{2}$ to 6 feet, 62 to 74 inches ($n = 10$), 60 to 81 inches ($n = 50$).

The range is useful. It reveals, without a graph, the extent of the distribution. But it doesn't tell us anything about the form of the distribution; it doesn't show us how the frequencies of different occurrences are distributed. But if the distribution is known to be normal, then with the mean and what is called the **standard deviation** we have all we need to completely describe its form.

We can tell just how a sample or population is normally distributed if we have the means and standard deviations. So if you consider the relative frequencies of different occurrences in a distribution as a direct indication of the probabilities of their occurring, then you have got yourself a very useful tool for statistical inference.

The standard deviation is the square root of the average squared deviations from the mean (called the root mean square). Each individual number (x) deviates a certain amount ($x - \bar{x}$) from the mean—sometimes the deviation may be zero. To get the standard deviation, square each of those deviations, add them together, divide by the number of them, and then take the square root. In short, using the Greek small sigma, σ (our s) for the population standard deviation, and the Greek large sigma Σ for summation:

$$\sigma = \sqrt{\frac{\Sigma(x - \mu)^2}{N}}$$

and for the sample:

$$s = \sqrt{\frac{\Sigma(x - \bar{x})^2}{n}}$$

It is important to recognize that the mean and standard deviation of a normal distribution completely describe it. The mean locates the central axis; the standard deviation reveals the spread and shape.

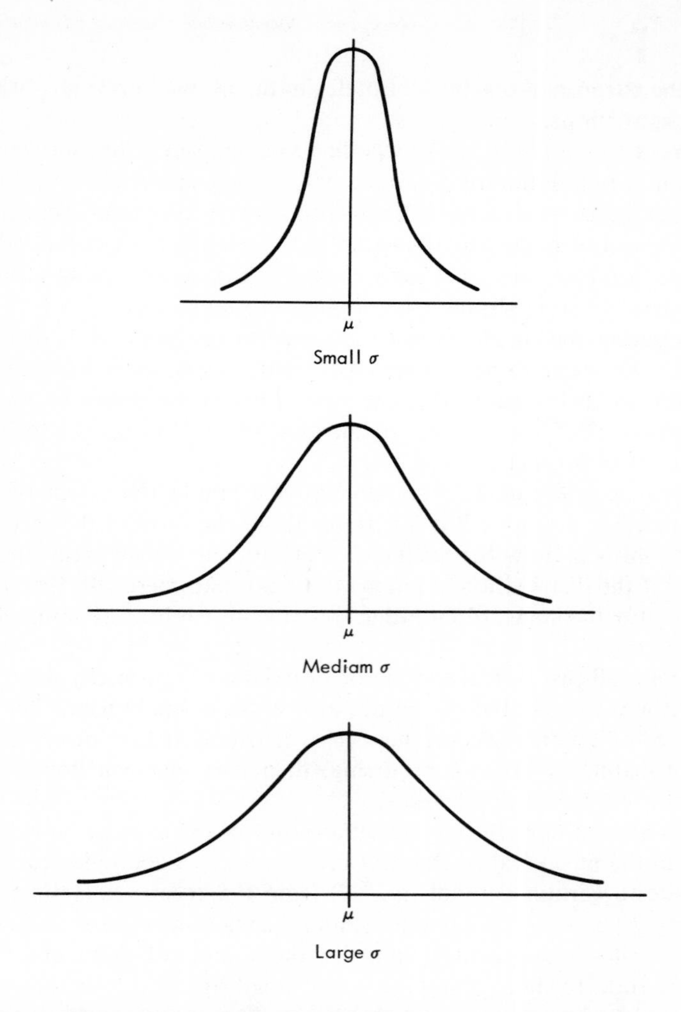

5–5 THE STANDARD NORMAL DISTRIBUTION (z TABLE)

We will often want to find the probability that a random sample will fall in (a) the area between the mean and some value of x, (b) below or above some value of x, or (c) between two values of x (call them x_1 and x_2).

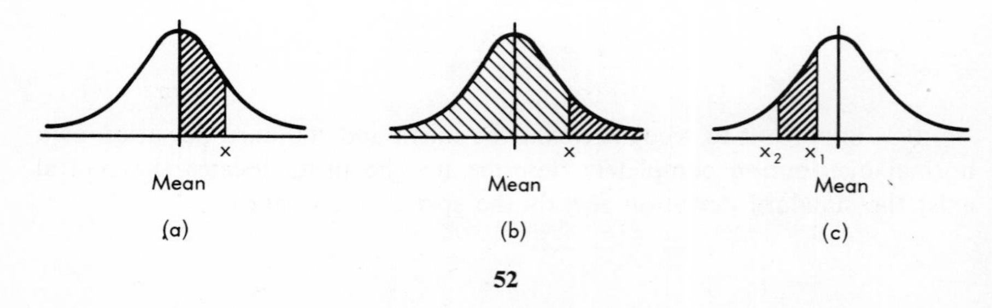

Given the mean and standard deviation of any normal distribution, we can find the probabilities related with any z by using Table IV(a) or Table IV(b) on pages 236 and 237 in the appendix. Three things to remember: z is the number of standard deviations of x from the mean (z *at* the mean is 0); plus z's are to the right of the mean, minus z's to the left; and the distribution is symmetrical, making the probabilities involving z on the right side of the mean ($+z$) the same as those involving the same z on the left side of the mean ($-z$).

Suppose we have a population distribution with $\mu = 100, \sigma = 13$, and $x = 122$. What is the probability that a random variable from the population will fall between the mean and x? How many standard deviations is the x from μ?

$$z = \frac{x - \mu}{\sigma} = \frac{122 - 100}{13} = \frac{22}{13} = 1.69$$

In Table IV(a), go down the z column to 1.6. Go across from there to the column headed 0.09 ($1.6 + 0.09 = 1.69$). The probability that a variable chosen at random will lie between μ and x is 0.4545. The probability that a random sample will lie beyond x is $0.5000 - 0.4545 = 0.0455$—or see Table IV(b) and get 0.0455.

In common use are the probabilities that a random variable will fall between $\mu + 1\sigma$, $\mu + 2\sigma$, and $\mu + 3\sigma$. These turn out to be 0.3413, 0.4772, and 0.4987, respectively. In even more common use are the probabilities that a random variable will fall in the area between μ plus and minus 1σ, and μ plus and minus 2σ, and μ plus and minus 3σ.

$$\mu \pm 1\sigma \text{ gives } 0.3413 + 0.3413 = 0.6826$$
$$\mu \pm 2\sigma \text{ gives } 0.4772 + 0.4772 = 0.9544$$
$$\mu \pm 3\sigma \text{ gives } 0.4987 + 0.4987 = 0.9974$$

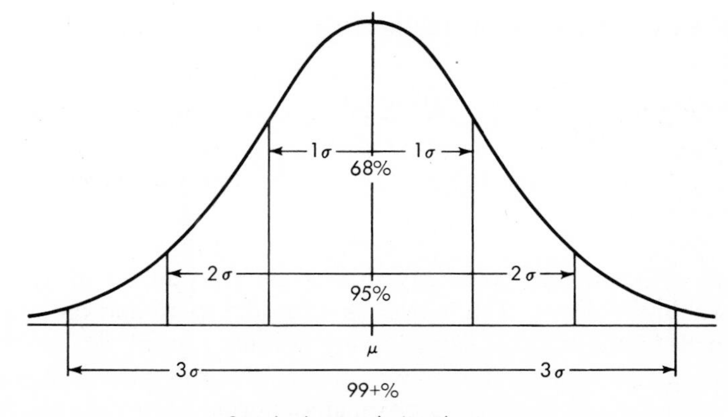

Standard normal distribution

This important simplification may be developed. Between the mean and plus and minus 1σ gives 68% of the whole; plus and minus 2σ gives 95%; plus and minus 3σ gives 99^{+}% (practically all of the whole).

IQ Scores

Theoretically, the mean of the population of IQs is 100. If yours is above that, you may be better than average. If it's below that, well, the sample was probably biased.

Anyway, the standard deviation of IQs across the United States is about 13. Assume a normal distribution (psychological testing bears out that this is a pretty fair assumption).

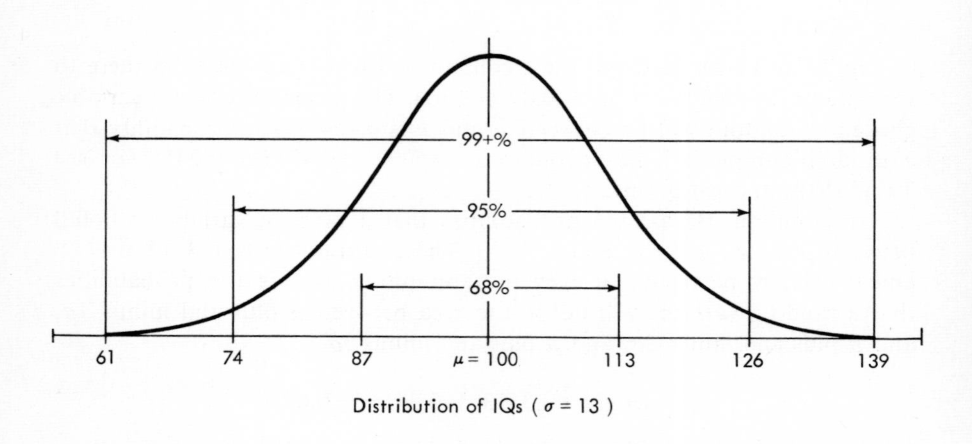

Distribution of IQs (σ = 13)

Making use of what we have already calculated, we can say:

About 68% of the people fall in the $\mu \pm 1\sigma$ interval, which is 100 ± 1 (13) or 87 to 113.

About 95% of the people fall in the $\mu \pm 2\sigma$ interval, which is 100 ± 2 (13) or 74 to 126.

99^{+}% of the people fall in the $\mu \pm 3\sigma$ interval, which is 100 ± 3 (13) or 61 to 139.

When someone tells you, "My sister Agnes has an IQ of 154," you'd better swallow it slowly. That's over 4σ from the mean and the probability is too small to find in Table IV(a). A very unusual Agnes!

There are all sorts of things we can do with this new statistical tool, as long as the distribution is approximately normal. For instance, in some public schools, students who show an IQ of 80 or below are eligible for "special classes." What percent of the students would you expect to have IQs of 80 or below?

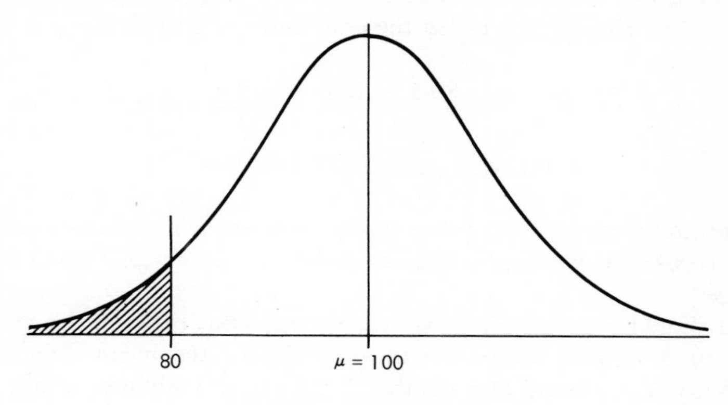

Distribution of IQs (σ = 13)

It is that part of the distribution 80 or below that we want.
How many standard deviations is it to 80?

$$z = \frac{x - \mu}{\sigma} = \frac{80 - 100}{13} = \frac{-20}{13} = -1.54$$

80 is -1.54 standard deviations from μ. The minus indicates only that x is to the left of μ.

Turn to Table IV(b) in the appendix again; it shows the probabilities above $+z$ or below $-z$, saving us the subtraction of Table IV(a) values from 0.5000.

Using $z = 1.54$, we find 0.0618:

$$P_{(80 \text{ or below})} = 0.062$$

What is the probability of a student having an IQ of 80 (no more, no less)? The line at 80 has no area, so we use 79.5 to 80.5.

$$z_{79.5} = \frac{79.5 - 100}{13} = \frac{-20.5}{13} = -1.58$$

$$z_{80.5} = \frac{80.5 - 100}{13} = \frac{-19.5}{13} = -1.50$$

$$P_{(80, \text{ no more, no less})} = 0.4429 - 0.4332 = 0.0097 = 0.010$$

The distribution of IQs is not continuous; you probably never heard of a person having an IQ of 80.23. IQ scores are usually considered to be discrete; they go in whole number jumps: 79, 80, 81. . . .

Looking at it this way, the area 80 or below might better be considered the area below 80.5 (this is called the continuity adjustment).

$$z = \frac{x - \mu}{\sigma} = \frac{80.5 - 100}{13} = \frac{-19.5}{13} = -1.5$$

$$P_{(80.5 \text{ or below})} = 0.0668 = 0.067$$

The continuity adjustment didn't really make much difference (0.067 − 0.062 = 0.005). On what general areas under the curve will it make the most difference?

You should now be able to get the probabilities between the mean and z standard deviations above the mean or below the mean directly from Table IV(a). You should also be able to get the probabilities above $+z$ or probabilities below $-z$ directly from Table IV(b). Probabilities between two z's you should get by subtracting probabilities when z's are on the same side of the mean, or by adding when z's are on opposite sides of the mean.

Case I: z's both on same side of mean

$$P_{(\text{between 110 and 120 inclusive})} = ?$$

Bringing in the continuity adjustment would change the interval to 109.5 to 120.5.

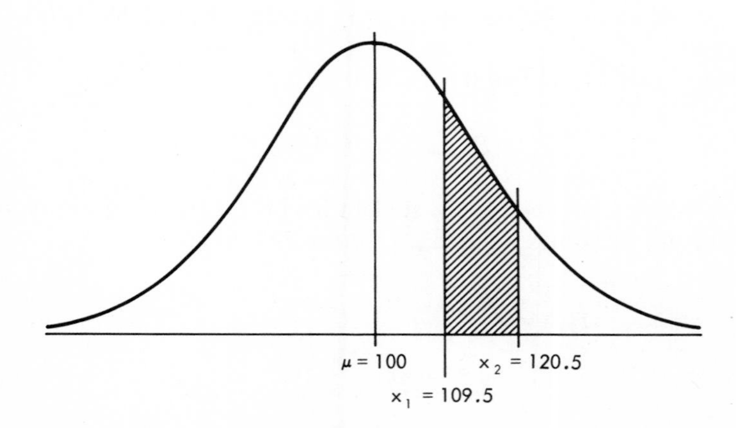

Probability distribution for IQs

$$x_1 = 110$$

$$z_1 = \frac{x_1 - \mu}{\sigma} = \frac{109.5 - 100}{13} = \frac{9.5}{13} = 0.731 = 0.73$$

$$P_{(\text{between } \mu \text{ and } 110)} = 0.2673$$

$$x_2 = 120$$

$$z_2 = \frac{x_2 - \mu}{\sigma} = \frac{120.5 - 100}{13} = \frac{20.5}{13} = 1.577 = 1.58$$

$$P_{(\text{between } \mu \text{ and } 120)} = 0.4429$$

$$P_{(\text{between } 110 \text{ and } 120)} = 0.4429 - 0.2673 = 0.1756 = 0.176$$

Had we not used the continuity adjustment, we would have got 0.159, so this is our justification for using it.

Case II: z's on opposite sides of the mean

$$P_{(\text{between } 90 \text{ and } 120 \text{ inclusive})} = ?$$

Bringing in the continuity adjustment again would change the interval to 89.5 to 120.5.

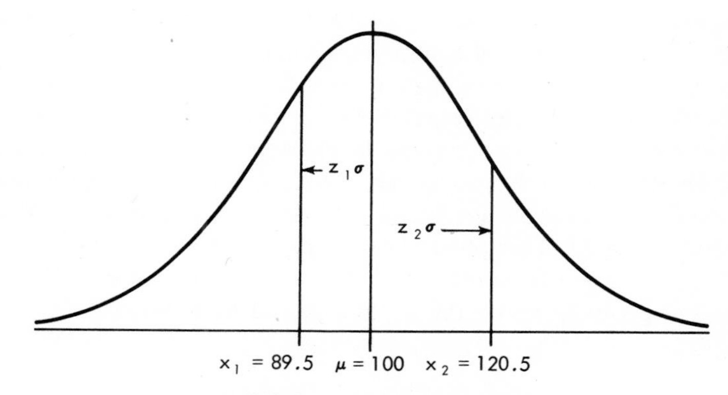

Probability distribution for IQs

$$x_1 = 90$$

$$z_1 = \frac{x_1 - \mu}{\sigma} = \frac{89.5 - 100}{13} = \frac{-10.5}{13} = 0.808 = 0.81$$

$$x_2 = 120$$

$$z_2 = 1.58$$

$$P_{(\text{between } 89.5 \text{ and } 120.5)} = 0.2910 + 0.4429 = 0.7339 = 0.734$$

(Using 90 and 120 unadjusted would have brought us 0.718. Is the difference worth it?)

So, do all the exercises at the beginning of this chapter; it will put you in good shape for what's coming. Do some of them over again and you'll be in even better shape.

5-6 SUMMARY

It is evident that our statistical inference is dependent upon probability distributions.

The first probability distribution we examined was the binomial distribution where a happening could happen in only one of two ways and the individual probability of that happening remained constant. The distribution showed the probabilities of different numbers of these happenings.

Next we looked at a normal distribution, shown by the bell-shaped curve, where the highest probabilities appear near the mean, then decrease rapidly, and taper off indefinitely on the tails. Many natural phenomena have probabilities quite normally distributed. Also, some important sample statistics have normal distributions. Even the binomial often may be considered normal, as we shall see in Chapter 7.

A normal distribution can be completely described by its mean and standard deviation. Given the number of standard deviations (z) that a variable (x) is from the mean, we can tell, from a standard normal table, the probabilities of variables being above x, below x, or between x_1 and x_2.

In the next chapter we are going to practice some of the computations you will be involved in as you go on. This attention to computations and fundamental arithmetic is not because I have any desire to dwell on the computational aspect; I simply don't want such stuff to bog you down later. Too often students get pretty excited about the promises of statistical inference but get hung up on the arithmetic and fade away.

Finally,

Sample	**Population**
$\bar{x} = \frac{\Sigma x}{n}$	$\mu = \frac{\Sigma x}{N}$
$s = \sqrt{\frac{\Sigma(x - \bar{x})^2}{n}}$	$\sigma = \sqrt{\frac{\Sigma(x - \mu)^2}{N}}$
$z = \frac{x - \bar{x}}{s}$	$z = \frac{x - \mu}{\sigma}$

Later on we usually substitute s for σ and change n to $n - 1$. Our formula for using s for σ becomes:

$$s = \sqrt{\frac{\Sigma(x - \bar{x})^2}{n - 1}}$$

More on this when the time comes.

• 6 •

Computations and Calculations

Man has been on earth 5,000,010 years. I know this is so because it was 5,000,000 years 10 years ago.

It is a common mistake to think that extensive mathematical calculations insure the accuracy of one's conclusions.

OBJECTIVES

To simplify the arithmetic of the decision process by coming up with some guidelines as to how far to carry out computations.

6–1 EXERCISES

1. The numbers of onions in bunches were counted with these results: 19, 23, 22, 19, 23, 24, 17, 22, 22, 23. Find Σx. Do you get 214? What is the precision of this result?

2. Find the mean number of onions per bunch. We intend to use this mean in computing the standard deviation. So what precision do you suggest we maintain? 21.4? Why?

3. Find the deviation of each count from the mean count (for instance, $19 - 21.4 = -2.4$). What determines the precision of these differences? Remember, this is only one step in our whole calculation.

4. Square these deviations—for instance, $(-2.4)^2 = 5.76$. Now it's a matter of significant figures again. You use one more significant figure in the square of the number than in the original number because you have more calculating to do. Next, be sure that all the addends have the same precision. This should get you a sum of 46.40.

5. Now try $s = \sqrt{\Sigma(x - \bar{x})^2/(n - 1)}$. Dividing by 9 (an exact number) brings us out with still four significant figures. The limitations of our square root table being three significant figures, we should round off $\sqrt{5.156}$ to $\sqrt{5.16}$ and then get 2.272.

6. Last, we want $\bar{x} \pm 3s$, so $21.4 \pm 3(2.272) = 21.4 \pm 6.816 = 21.4 \pm 6.8$. For adding and subtracting, round off 6.81 to 6.8 to get the same precision as 21.4. As a final result this should give, rounded off to the *original* precision, 15 to 28 onions per bunch, a fairly safe estimate.

   Recognize again that these are only arbitrary rules or suggestions that we have followed. Different people are likely to interpret them differently and emerge with slightly different answers.

7. The contents of 10 packages of lettuce seeds weighed the following number of grams: 1.9, 2.3, 1.6, 1.7, 1.7, 2.2, 2.0, 2.2, 1.9, 2.4. Find Σx. What is the precision of this sum? Did you get 19.9 grams?

8. Find the mean weight per package. How many significant figures? Why?

9. Find each deviation from your mean and square it. For instance,

$$\begin{aligned}(x - \bar{x})^2 &= (2.0 - 1.99)^2 = (0.01)^2 \\ &= 0.0001\end{aligned}$$

and $$(2.3 - 1.99)^2 = (0.31)^2 = 0.0961$$

10. In finding the sum of the squared deviations, you really don't have any good grounds for using more decimal places than the one you added a while back to the original precision of one decimal place. If you keep four decimal places, the sum is 0.6890; if you round each addend to two decimal places, the sum is 0.68.

11. $s = \sqrt{\dfrac{\Sigma(x - \bar{x})^2}{n - 1}} = \sqrt{\dfrac{0.68}{9}} = \sqrt{0.0756} = 0.27$

12. Try for $\mu \pm 3s$ to one decimal place. Then $\mu = 1.99$ rounds off to 2.0. And $\mu \pm 3s = 2.0 \pm 3(0.3) = 2.0 \pm 0.9 = 1.1$ to 2.9 grams.

13. If $\pi = 0.436$, the population proportion of cancer victims having group A blood, what is the probability that 30 or more cancer victims will have group A in a sample of 50? Use continuity adjustment.

14. In human shivering, the mean of the oscillating rhythmic muscle tremors is 15.0 tremors per second with a standard deviation of 1.70 tremors. What is the likelihood that a shivering man will have more than 20.0 tremors per second if the distribution of tremors is normal?

15. Long study of the rocks in glacial till in a well-defined New England area shows 78% to be some variety of granite. In a random sample of 100 rocks from that area, what is the probability that 50 or more will be granite?

16. Select 100 consecutive two-digit numbers from the random number table. Using the intervals 1–9, 10–18, 19–27, . . . , 91–99, make a frequency distribution of the results. First, though, what sort of a shape do you expect to get, approximately?

17. In 1968 the American railways owned 208,111 miles of track, 63 miles being built that year. What percent of the track mileage was newly constructed? What determines the number of significant figures in your answer?

18. The temperature is 70° F. What does that mean from an interval point of view?

19. When you say his body temperature is 97.8°F, then what are the limits of this interval?

20. We want to pour cement for a garage floor 18 ft × 23 ft. If these measurements are to the nearest foot, find the floor area, using the interval lower boundaries and then the upper boundaries. How much do they differ? Does this suggest a justification for rounding off the area to two significant figures, the number given?

6–2 EXACT NUMBERS

There are two common computational dangers (besides making mistakes): (1) that you get involved in an overpowering computational activity the results of which may be much less significant than all those figures and decimal places suggest and (2) that you go rounding off figures and decimal places that could have been meaningful in the decision process. To determine how far to carry out computations with confidence, you must recognize the difference between exact and approximate numbers.

Exact numbers are always whole numbers. There are no parts. It is 1, 2, 3, . . . and nothing in between. The count is exact: 13, 1032, 106, 729, ... The counts may be wrong, but they are represented by exact numbers, no more, no less.

Now for fundamental operations with exact numbers.

Addition:

```
 4313
   27
 ----
 4340
```

No question about that.

Subtraction:

```
 4313
   27
 ----
 4286
```

No question about this either.

Multiplication:

```
   4313
     27
 -------
  30191
  8626
 -------
 116,451
```

Still no problem.

Now for division: $\require{enclose} 27\enclose{longdiv}{4313.}\kern-.2ex$ $= 159.7407\ldots$

Here is a problem. Given 4313 apples in 27 baskets, what is the mean number of apples per basket?

$$\mu = \frac{\Sigma x}{N} = \frac{4313}{27} = 159.7407\ldots$$

Don't scoff. This is a kind of computational decision you will be having to make continually. How many places shall you carry out the quotient from the division of two exact numbers?

This suggestion may help: If the result is final, round off to the nearest whole number; if the result is to be used in further calculations, keep one decimal place.

If you are after the mean for its own sake:

$$\mu = \frac{\Sigma x}{N} = \frac{4313}{27} = 159.74 \cdots = 160 \text{ apples}$$

If you are using the mean to compute the standard deviation:

$$\sigma = \sqrt{\frac{\Sigma(x - \mu)^2}{N}} = \sqrt{\frac{\Sigma(x - 159.7)^2}{27}}$$

If $\Sigma(x - 159.7)^2$ equals 158.8, then

$$\sqrt{\frac{158.8}{27}} = \sqrt{5.88} = ?$$

I am trying to condense my suggestions. You may have to improvise, depending on what sort of arithmetical tangle you wind up in. But have courage; you are really trying to look in the eye situations that are commonly ignored.

I should remind you that in rounding off you have to decide whether what you drop warrants leaving the remaining last digit as it is or whether it raises it by 1 ($21.86 = 21.9$, $21.83 = 21.8$). Furthermore, if what you drop is a single 5, it is good practice to raise your remaining last digit by 1 if, and only if, it is an odd number ($21.85 = 21.8$, $21.75 = 21.8$). This odd number business tends to reduce an error you'd be building up if you always added 1 each time you round off from a 5.

6–3 SQUARE ROOTS

$$\sqrt{9} = 3$$
$$\sqrt{90} = 9.487\ldots$$
$$\sqrt{900} = 30$$
$$\sqrt{9000} = 94.87\ldots$$
$$\sqrt{90000} = 300$$
$$\sqrt{900000} = 948.7\ldots$$

For every sequence of digits, wherever the decimal point, there are two possible sequences of digits for their square roots.

$$\sqrt{5.23} = 2.287\ldots$$
$$\sqrt{52.3} = 7.232\ldots$$
$$\sqrt{523.0} = 22.87\ldots$$
$$\sqrt{5230.0} = 72.32\ldots, \text{etc.}$$

You have to decide which to use and where to put the decimal point. You could use the coupling method. Go from the decimal point to either side by coupling.

$$\sqrt{\overset{\frown}{53}\overset{\frown}{20}}$$
$$\sqrt{\overset{\frown}{05}\overset{\frown}{32}}$$
$$\sqrt{0.\overset{\frown}{53}\overset{\frown}{20}}$$
$$\sqrt{0.\overset{\frown}{05}\overset{\frown}{32}}$$
$$\sqrt{\overset{\frown}{53}.\overset{\frown}{20}}$$
$$\sqrt{\overset{\frown}{05}.\overset{\frown}{32}}$$

Use a zero to fill out any incomplete couple.

It is the first complete or incomplete couple that gives you the first digit of your square root. What is the largest whole number that when squared does not exceed the first couple? This becomes the first digit of your square root. *And* each couple represents one decimal place to the left or right of your first digit.

$$\sqrt{5320} = 7X.XX$$
$$\sqrt{0532} = 2X.XX$$
$$\sqrt{0.5320} = 0.7XXX$$
$$\sqrt{0.0532} = 0.2XXX$$
$$\sqrt{53.20} = 7.XXX$$
$$\sqrt{05.32} = 2.XXX$$

Now go to your square root table (Table III in the appendix). The digits following 532 are 2307 and 7294, the first four digits of the two possible sequences.

This will give you

$$\sqrt{5320} = 7X.XX \quad = 72.94$$
$$\sqrt{0532} = 2X.XX \quad = 23.07$$
$$\sqrt{0.5320} = 0.7XXX = 0.7294$$
$$\sqrt{0.0532} = 0.2XXX = 0.2307$$
$$\sqrt{53.20} = 7.XXX \quad = 7.294$$
$$\sqrt{05.32} = 2.XXX \quad = 2.307$$

Study these results. See how careful coupling will always reveal the first digit of your square root and its relation to the decimal point; your square root table shows the other digits. Notice that square roots are given for three-digit numbers only; but the square roots themselves are given as four-digit numbers.

Now go back for the standard deviation of apple counts:

$$\sigma = \sqrt{\frac{\Sigma(x - \mu)^2}{N}}$$
$$= \sqrt{\frac{158.8}{27}} = \sqrt{5.88} = 2.425 \text{ or } 2.42$$

Round off to three digits the number of digits you have under the square root sign.

6–4 APPROXIMATE NUMBERS

All measurements are approximate; with a more precise measuring device, we can always make finer and finer measurements.

Suppose we measure the lengths of 2-inch brads to the hundredth of an inch. We could put these measurements on a continuous horizontal axis. Assume a normal distribution with $3\sigma = 0.25$ inches.

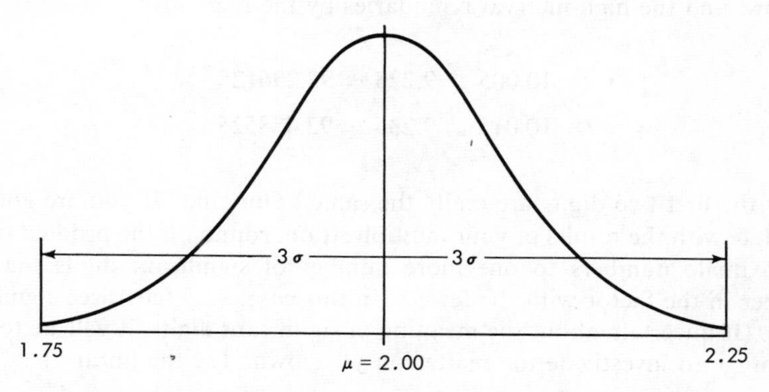

Probability distribution for length in inches of 2-inch brads

The length of a brad will probably fall between $\mu \pm 3\sigma$; it can have any value whatever in that interval. Suppose we are given three brad lengths to a precision of $\frac{1}{100}$ of an inch.

Addition:

$$\begin{array}{r} 2.23 \\ 1.98 \\ 2.04 \\ \hline 6.25 \end{array}$$

All the elements should be given to the same precision (same number of decimal places). The sum should have the same precision.

Subtraction:

$$\begin{array}{r} 2.23 \\ 1.98 \\ \hline .25 \end{array}$$

Both elements should be given to the same precision (the same number of decimal places). The difference should have the same precision.

Multiplication:

$$\begin{array}{r} 10.01 \\ 9.23 \\ \hline 3003 \\ 2002 \\ 9009 \\ \hline 92.3923 \end{array}$$

This situation is different because we are multiplying an approximate number by an approximate number.

10.01 means anywhere between 10.005 and 10.015, and 9.23 means anywhere between 9.225 and 9.235. Multiplying the low-interval boundaries by the low, and the high-interval boundaries by the high, gives

$$10.005 \times 9.225 = 92.296125$$
$$10.015 \times 9.235 = 92.488525$$

(Only the first two digits are really the same.) Our rule: If you are going to calculate with the results of your multiplication, round off the product of two approximate numbers to one more number of significant digits than the number in the factor with the fewest; in this case, 9.23 has three significant digits. (If uncertain about the meaning of significant digits, it will be to your advantage to investigate the matter on your own. Try the library.)

$$10.01 \times 9.23 = 92.3923 = 92.39$$

If the product is a final result, keep only the number of significant figures in the factor with the fewest (9.23, so 92.4).

Division: $9.23 \div 2.36 = 3.9110168\ldots$

The same rule holds as for multiplication: Round off to one more than the least number of significant figures given if the results are to be used in calculations. Use the same number of significant digits as in the one with the least number of significant digits, if results are final.

To be used: $9.23 \div 2.36 = 3.911$

Final result: $9.23 \div 2.36 = 3.91$

This amounts, in both multiplication and division, to carrying along an extra significant figure and rounding off when you get through.

The same suggestion helps with square roots

$$\sqrt{3.911} = \sqrt{3.91} = 1.977$$

if you are going to use it in calculations, or 1.98 if this is your final result.

6–5 CONTINUITY ADJUSTMENT

When the probability questions involve exact numbers, you should usually make an interval on your graph.

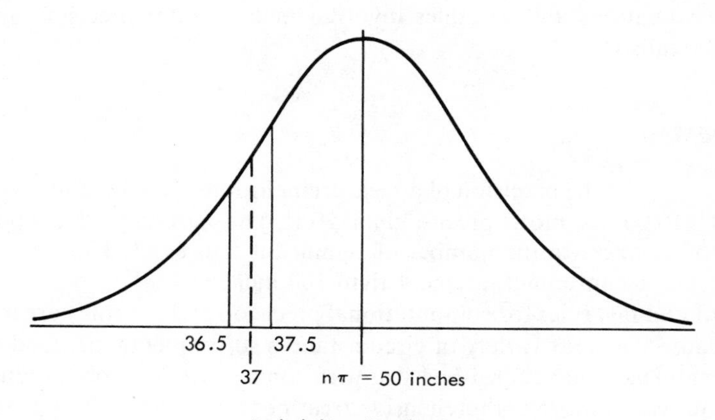

Probability distribution

The vertical line at 37, of course, has no area. To make an interval, call the exact number 37, the interval 36.5 to 37.5. Now you have an area and can find the probability of 37 by finding what proportion the area above your interval is of the whole area under the curve. When you want to find the probability of an exact number of results, you should usually go to intervals,

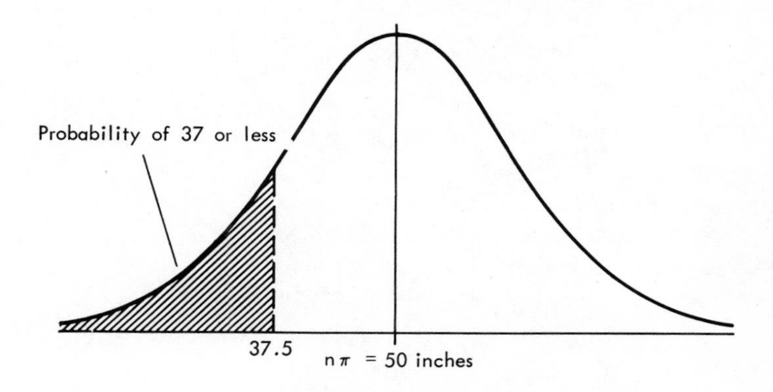

Probability distribution

With what may be considered continuous proportions and means, you don't usually need a continuity adjustment. If you want the probability of less than $p = 0.64$, don't change it to less than 0.635. If you want the probability of more than the sample mean 12.26, don't use 12.265. Use the continuity adjustment for exact numbers. In all cases, we should say "usually." You will have to use your judgment: Is greater precision and accuracy called for, or do the approximations and variables involved make greater precision and accuracy unrealistic?

6-6 SUMMARY

The precision of a measurement refers to the units of measurement; 10.001 is more precise than 10.1. The accuracy of a count or measurement refers to the number of significant figures; 113 inches is possibly a more accurate measurement than 100 inches.

Hard and fast rules for computational precision and accuracy are impractical because the great variety of circumstances so frequently demand exceptional treatment. I have furnished you with some guidelines, pretty much my own. However, a more comprehensive treatment of exact and approximate numbers would be a difficult and quite sophisticated project. But I wanted to call these matters to your attention and possibly increase your chances of handling precision and accuracy in your computations with some confidence.

Exact mathematics is mostly abstract. In life we are almost never exactly right, but don't allow that to let you scoff. Modern technology has got man to the moon and back by skillfully using thousands upon thousands of approximate numbers.

• 7 •

Mathematical Models

He was a poet and hated the approximate.

RAINER MARIA RILKE (1875–1926)
The Journal of My Other Self

OBJECTIVES

To look at the binomial distribution,
the normal distribution,
and Tchebycheff's theorem
in the light of their being
approximate mathematical models.

SYMBOLS AND FORMULAS FIRST USED IN THIS CHAPTER:

$\mu = n\pi$	the mean of sample counts
$\sigma = \sqrt{n\pi(1 - \pi)}$	the standard deviation of sample counts
$z = \dfrac{x - \mu}{\sigma}$	the number of standard deviations of x counts from the mean of the sample counts
k	the number of standard deviations from the mean of any distribution
$1 - \dfrac{1}{k^2}$	at least this proportion of any distribution within $\mu \pm k\sigma$ (Tchebycheff)

7-1 EXERCISES

1. What conditions allow using the binomial tables for finding probabilities?

2. What conditions allow using a normal distribution as an approximation of the binomial? When would you use it? What information do you need to use it?

3. Cite a distribution that is not binomial. Explain why it is not binomial.

4. Under what circumstances would you employ Tchebycheff's theorem? What information do you need to use it?

5. A traffic light at a particular intersection is red 0.50 of the time, green 0.50 of the time, and never amber. A driver arriving randomly on his way to and from work records the color of the light 15 times. What is the probability that it was red exactly 7 times?

6. It is predicted that 40% of the remaining elm trees in a certain city will die next year. What could you say about the probability that exactly 4 of the 14 elms on Front Street will die?

7. In a poll of 50 randomly selected people in a city where 60% of the population supports the mayor's policies, what is the likelihood that less than 20 will be found to support the mayor?

8. It is estimated that 70% of the students in a college math class are listening to the instructor at any one time. What is the chance that all 15 in a class will be listening to the instructor at the same time?

9. On an American roulette wheel there are 18 red compartments, 18 black compartments, and 2 compartments not colored. What is the probability of getting 8 reds (no more, no less) in 15 spins? The ball rolls, supposedly at random, into one compartment on every spin of the wheel.

10. The boiling point of water is measured in degrees centigrade by 50 people at a science laboratory. The boiling point of water is supposed to be 100°C. Assuming that the water was pure and the atmospheric pressure correct for sea level, the difference in measurements must have been due only to chance error. Assuming a normal distribution with $\sigma = 0.015$, sketch a graph of the distribution of chance errors, locating the 1, 2, and 3σ's.

Early application of the normal curve to variations in astronomical measurements earned it the name *normal curve of errors*. It is also called the Gaussian curve after the mathematician Karl Gauss, who first applied it to such errors early in the nineteenth century.

11. Suppose that the heights of 100 grass blades are measured on a lawn cut the week before, giving a mean of 1.5 inches with a standard deviation of 0.30 inches. What is the probability that a randomly selected single grass blade will exceed 2.9 inches? Assume a normal distribution.

12. According to Tchebycheff's theorem, what is the smallest proportion of a distribution of measurements or counts that can lie within 4 standard deviations of the mean?

13. The beaks of 77 bluejays were measured. $x = 2.20$ centimeters, $s = 0.15$ centimeters. What is your estimate of the least proportion of the bluejays that had beaks between 1.90 and 2.50 centimeters?

14. Given 2.23, 4.6, 3.721, 5.8, and 2.7682 centimeters as measurements of five things, find what you consider to be an appropriate mean.

15. Find the standard deviation of the population of the five measurements in the previous exercise.

16. The bathtub curve for electrical motor life after repair showed a mean of 1100 hours of service with a standard deviation of 125 hours. What is the probability that a motor randomly selected will last between 850 and 1350 hours. Should you use a continuity adjustment?

7–2 THE BINOMIAL PROBABILITY DISTRIBUTION

The binomial table in our appendix shows the probabilities for different counts (x) for different individual probabilities (π) for different sample sizes (n) 2 through 15. (Why not $n = 1$?) But the table at $n = 15$ is beginning to spread out. Run n up to 100 and the tables begin to get too heavy to carry.

7–3 THE NORMAL DISTRIBUTION FOR THE BINOMIAL

If you can find a mathematical model that pretty well describes a life situation, like $A = lw$ for the area of a rectangle, or $E = mc^2$ (Einstein's famous energy formula), then you can express the consequences with numbers instead of with descriptive adjectives. Describing a house lot as large is obviously imprecise compared with saying it is 2.5 acres.

Saying that an event is very likely is not nearly as revealing as saying that its probability is 0.893. But in order to come up with any such numerical index of probability we usually need a mathematical model to work from. This we do have in the binomial theorem (binomial expansion) where any particular term is $\binom{n}{x}\pi x(1-\pi)^{n-x}$. The algebra and combinatorics involved are fascinating, but we are pursuing a general approach to statistical inference and are not going to get too involved in mathematical models. It is enough for us to know that the binomial expansion is the mathematical model from which the binomial probabilities were calculated.

But as the sample size increases, the binomial tables become more and more bulky. So we look around for a different mathematical model to help us out. And sure enough (Galton's ghost!), here comes the normal distribution again. Mathematicians assure us that as the sample size increases, the binomial distribution gets to look more and more like a normal curve. When π differs much from 0.500, then the binomial gets to be skewed (lopsided) in one direction or the other (as the histograms in Chapter 5 showed). Happily, though, as the sample size increases, the lopsidedness becomes less and less evident, and under the right circumstances we can still approximate the binomial with a normal distribution.

Under the right circumstances? Here is a handy rule of thumb to help you decide whether it is safe to use the normal distribution: both $n\pi$ and $n(1-\pi)$ should be greater than 5. (Long ago, a brewer would stick his thumb into his vat to check the temperature of the brew: hence, *rule of thumb.*)

To use the z table with this approximation, however, you will need a mean and a standard deviation. Coming right up! $\mu = n\pi$; $\sigma = \sqrt{n\pi(1-\pi)}$. It is fairly obvious, in the first case, that the mean or most likely count will be π times the sample size. What's the most likely number of heads in 100 tosses of a coin? Right! $\mu = n\pi = 100(0.500) = 50$. (The standard deviation is not quite so easy to see intuitively.) But we've got here the tools we need for solving more problems of the binomial type.

Bottles of Soured Milk

What is the probability that if the store manager checks 100 leftover quarts of milk at his corner market early Monday morning, one of them will have "turned"? Well, what is π? Checking from way back, the proportion of turned quarts in this particular store on Monday mornings appears to be about 10%. (Unrealistically high.)

This presents us with $\pi = 0.10$, $n = 100$, $x = 1$. As $n = 100$ is way beyond our binomial tables, we approximate the binomial distribution with a normal distribution.

$$\mu = n\pi = (100)(0.10) = 10 \text{ quarts}$$
$$\begin{aligned}\sigma &= \sqrt{n\pi(1-\pi)} \\ &= \sqrt{100(0.10)(0.90)} = \sqrt{100(0.09)} \\ &= \sqrt{9} = 3 \text{ quarts}\end{aligned}$$

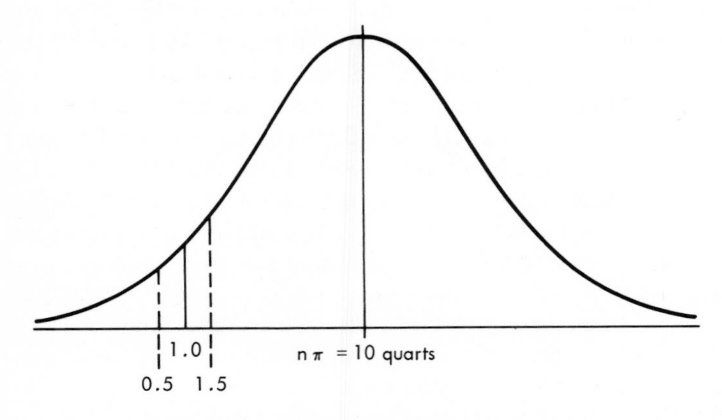

Distribution of number of "turned" quarts of milk
(n = 100, σ = 3 quarts)

You see at once that 1, as it is, has no area and hence shows no probability. We shall have to compensate for this lack of area by calling 1 an interval from 0.5 to 1.5. Now it has an area. This is the continuity adjustment, again.

$$z_{0.5} = \frac{x-\mu}{\sigma} = \frac{0.5-10}{3} = \frac{-9.5}{3} = -3.17$$

$$z_{1.5} = \frac{x-\mu}{\sigma} = \frac{1.5-10}{3} = \frac{-8.5}{3} = -2.83$$

Area for $z = -3.17$ is 0.4992 (not from *your* table)

Area for $z = -2.83$ is 0.4977

$$P_{(\text{exactly } 1)} = 0.4992 - 0.4977 = 0.0015 = 0.002$$

But is π too far from 0.500 and is n too small? That is the question. So we come up with that rule of thumb which handles both sides of the question at once, requiring that $n\pi$ and also $n(1-\pi)$ be greater than 5.

$$n\pi = [100(0.10)] = 10 > 5$$
$$n(1 - \pi) = [100(0.90)] = 90 > 5$$

Both are greater than 5. Good! We'll accept the result. But really we could have wasted a lot of time by not using this rule before we started calculations. One or both of our deciders might have been 5 or less, and that would have invalidated all our calculations. Check $n\pi$ and $n(1 - \pi)$ *first*!

Incidentally, these deciders are not God's laws nor even mathematical truths. They are arbitrary, expedient rules for human beings, designed to make the normal distribution a safer mathematical model for the binomial distribution.

School Committee and Superintendent

It is reported to the school committee that one-half of the faculty members in a certain school district think the superintendent is doing a poor job. A check is made of 14 randomly selected faculty members. What is the probability that 7 of them (no more, no less) will say the superintendent is doing a poor job, if the report about half the faculty thinking so is really true?

Before calculating, try guessing. If half the faculty members think the superintendent is doing a poor job, what is the probability that half a sample of 14 faculty members will? Really think about it. It becomes clear that 7 out of 14 is certainly not a sure thing; in fact, the number in the sample who think he's doing a poor job could be anywhere from 0 to 14, inclusive. But how sure is 7? Isn't it surprising that common sense will not answer this simple problem? Even some sustained thinking won't do it. Elementary probability has some strange faces.

Try the binomial probability table:

$$\pi = 0.500$$
$$n = 14$$
$$x = 7$$
$$P_{(\text{exactly } 7)} = 0.209 \text{ or } 0.21$$

The number 7 will turn up about $\frac{1}{5}$ of the time if samples of 14 are randomly selected.

But suppose we didn't have a binomial probability table. Could we have done the problem? Using our rule of thumb, we find $n\pi = 14(0.5) = 7$, and $n(1 - \pi) = 14(0.5) = 7$. Both are more than 5. This will allow us to use the normal approximation. Let's see what happens.

$$\mu = n\pi = 14(0.500) = 7$$

$$\sigma = \sqrt{n\pi(1-\pi)} = \sqrt{14(0.500)(1-0.500)}$$
$$= \sqrt{14(0.500)(0.500)} = \sqrt{14(0.250)}$$
$$= \sqrt{3.50} = 1.87$$

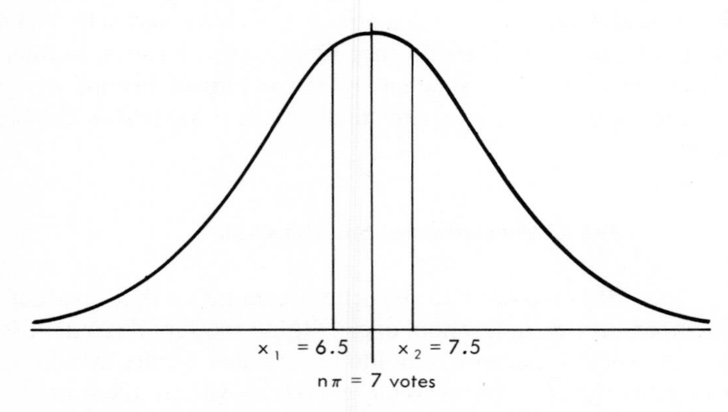

Distribution of votes about superintendent (n = 14, x = 7, σ = 1.87)

$$z_1 = \frac{x_1 - \mu}{\sigma} = \frac{6.5 - 7}{1.87} = -0.27$$

$$z_2 = \frac{x_2 - \mu}{\sigma} = \frac{7.5 - 7}{1.87} = 0.27$$

$$P_{(\text{exactly } 7)} = 0.1064 + 0.1064$$
$$= 0.2128 = 0.213 \text{ or } 0.21$$

You could say that it is a coincidence that this probability came out exactly the same as when we used the binomial table. Well, all right, but it is certainly no coincidence that they come out to be very close because that *is* the way the statistical ball bounces.

Cleveland Indians

A baseball player for the Cleveland Indians has a batting average of 0.300. (Consider this a lifetime average recorded up in baseball heaven ahead of time.) What is the probability he will get exactly 12 hits in the next 30 times at bat?

This is a binomial situation: he, each time, either does or doesn't; he hits or he doesn't. As $n = 30$, we can't look the probability up in our binomial probability table, since the table goes to $n = 15$ only.

Can we use the z table? Is the distribution approximately normal? Satisfactorily so, if $n\pi$ and $n(1 - \pi)$ are both more than 5.

$$\begin{aligned} n\pi &= 30(0.300) \\ &= 9 \\ n(1 - \pi) &= 30(1 - 0.300) \\ &= 30(0.700) \\ &= 21 \end{aligned}$$

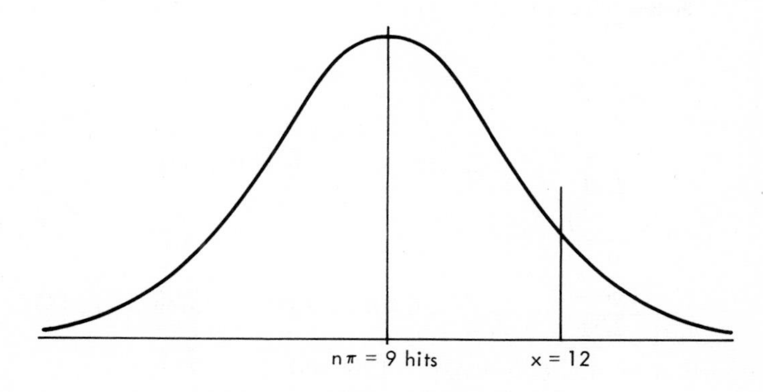

Distribution of ballplayer's number of hits (x)
in 30 tries (n) (π = 0.300)

Now, we can get over 12 or under 12. But we can't get *exactly* 12 because the vertical line at 12 doesn't have any area!

The trouble is we are using a continuous curve again, but we are talking about a discrete number of hits—in this case, 12. So we shall use the continuity adjustment and assume the area for exactly 12 to be from 11.5 to 12.5.

$$\begin{aligned} \sigma &= \sqrt{n\pi(1 - \pi)} \\ &= \sqrt{30(0.300)(0.700)} \\ &= \sqrt{30(0.210)} \\ &= \sqrt{6.30} \\ &= 2.51 \end{aligned}$$

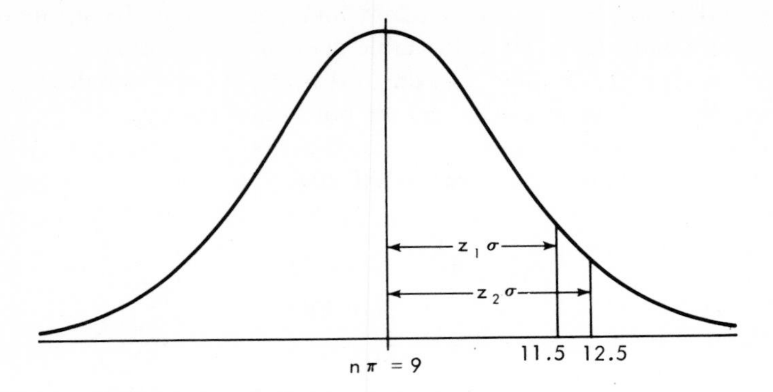

Probability distribution of hits ($\pi = 0.300, n = 30, \sigma = 2.51$)

$$z_2 = \frac{x_2 - \mu}{\sigma}$$

$$= \frac{12.5 - 9}{2.51} = \frac{3.5}{2.51} = 1.39 \quad \text{(area} = 0.4177)$$

$$z_1 = \frac{x_1 - \mu}{\sigma}$$

$$= \frac{11.5 - 9}{2.51} = \frac{2.5}{2.51} = 0.996 = 1.00 \qquad \text{(area} = 0.3413)$$

Proportion of area between $z = 1.39$ and $z = 1.00$ is

$$0.4177 - 0.3413 = 0.0764$$

$$P_{(12 \text{ hits in } 30 \text{ tries})} = 0.076$$

One handbook of tables for statisticians shows a binomial table for $n = 1$ through $n = 45$. (This takes 11 large pages.) Using this table the probability of $x = 12$ when $\pi = 0.300$ and $n = 30$ is 0.0749, which rounds off to 0.075.

Not bad! Our normal approximation gave us $P_{(x)} = 0.076$, which is off only 0.001 from what we would get using the binomial model.

Let's go at it from another angle. Suppose we ask the probability of *more* than 12 hits in the next 30 times at bat.

$$z_{12.5} = 1.39$$

See Table IV(b) in the appendix for area beyond $z_{12.5} = 0.0823$.

$$P_{(\text{over } 12 \text{ out of } 30)} = 0.082$$

If we had failed here to use this interval approach and used instead the single number 12:

$$z_{12} = \frac{x - \mu}{\sigma} = \frac{12 - 9}{2.51} = \frac{3}{2.51} = 1.20$$

$$P_{(\text{over } 12 \text{ in } 30)} = 0.1151 = 0.115$$

It does make a numerical difference. Whether it is considered a *significant* difference depends upon you and your problem. If our ballplayer was content to say "about 10% of the time," either 0.082 or 0.115 would furnish adequate grounds for that statement.

7–4 THE MIRACULOUS NORMAL CURVE

In these early problems we have asked questions, adding "assuming a normal distribution." Given a normal distribution and its mean and standard deviation, you can get the probability answers you need. And demanding a normal distribution, though a necessary condition, is not so much of a difficulty as you might think. The normal curve turns up again and again in mathematics and in life. It is like the ghost of Hamlet's father, *Hic et ubique*. Here and everywhere.

In industrial quality control work it continually amazed me to see how often raw distributions of measurements from out in the shop would assume a sort of bell shape when plotted in the lab. But that was Galton's rhapsody. Remember? Nature, man, machines, and mathematics yield approximate normal distributions with a miraculous frequency. Still, you mustn't depend on it without some reasonable assurance.

7–5 TCHEBYCHEFF'S THEOREM

Pafrutii L. Tchebycheff (died 1894) came up with an interesting theorem. If you think about it, you will realize that the computed standard deviation of any distribution, normal or not, is a sort of average of the deviations. You'd expect, then, that given the standard deviation, you ought to be able to come up with some sort of statement about the properties of the distribution.

This is just what Tchebycheff did. He proved mathematically that whatever the shape of the distribution, the least proportion of the population that would lie within k standard deviations from the population mean ($\mu \pm k\sigma$) could be expressed as the proportion $1 - (1/k^2)$. Use k here for the number of standard deviations. In dealing with the normal distribution, use z. In both cases, though, (k or z) we are referring to the number of standard deviations from the mean.

Tchebycheff's theorem seems fairly loose, not stating probabilities as precisely as we did with z for the normal distribution. Still, it does enable us to state something about any distribution as long as we know the mean and standard deviation.

Suppose, for instance, that we recorded the incomes of 10,000 people in a certain area. Then we computed the mean to be \$9000 and the standard deviation to be \$1000. What proportion of the salaries could we expect to lie within 1σ of the mean?

$$\mu = \$9000$$
$$\sigma = \$1000$$

Then, according to Tchebycheff, within 1σ of the mean will be at least

$$1 - \frac{1}{k^2} = 1 - \frac{1}{1^2} = 1 - 1 = 0$$

Not very enlightening! It says that whatever the distribution, *the least* proportion of it that we can possibly have between plus and minus one standard deviation from the mean is none. You knew that anyway; the least you can have *anywhere* is none.

But if $k = 2$, then at least

$$1 - \frac{1}{k^2} = 1 - \frac{1}{2^2} = 1 - \frac{1}{4} = \frac{3}{4}$$

At least $\frac{3}{4}$ of *any* distribution will lie between $\mu \pm 2\sigma$. So, with our sample of incomes, we know that at least $\frac{3}{4}$ of them will lie between \$7000 and \$11,000.

How about when $k = 3$? Then at least

$$1 - \frac{1}{k^2} = 1 - \frac{1}{3^2} = 1 - \frac{1}{9} = \frac{8}{9}$$

At least $\frac{8}{9}$ of any distribution will lie between $\mu \pm 3\sigma$. At least $\frac{8}{9}$ of the distribution of incomes will lie between \$6000 and \$12,000.

This is worth reconsidering, applying as it does to any distribution whatever.

Normal	**Tchebycheff**
$\mu \pm 1\sigma$—68%	$\mu \pm 1\sigma$—at least none
$\mu \pm 2\sigma$—95%	$\mu \pm 2\sigma$—at least $\frac{3}{4}$ or 75%
$\mu \pm 3\sigma$—99%	$\mu \pm 3\sigma$—at least $\frac{8}{9}$ or 89%

Tchebycheff's theorem is not as revealing as a normal distribution, but the concept does produce something specific and helpful in emergencies.

It is *very* easy to be careless and start calculating for z right off before thinking about the shape of the distribution. But watch out! If you are not *told* that it is a normal distribution, if you have no grounds for assuming it to be a normal distribution, then you'd better reach for Tchebycheff's theorem.

Also remember: without the mean and standard deviation, you can't apply either the normal distribution or Tchebycheff's theorem.

7–6 SUMMARY

When you have a probability distribution with the individual probability of an event (or its opposite) staying constant and you want to know the probability of a certain number of these events, use the binomial as a mathematical model.

If the binomial-distribution sample size is over the limit of your available tables, use a normal distribution as a model. As long as the sample gets larger and larger as π gets farther from 0.500, you will be reasonably safe with this approximate model. Guideline: require $n\pi$ to be greater than 5 and $n(1 - \pi)$ also to be greater than 5.

If you don't have a normal distribution but can get the mean and standard deviation, then you may use Tchebycheff's theorem as a mathematical model. The mean and the standard deviation will allow you to say something specific about any distribution.

The results of our arithmetical manipulations may not apply exactly as they did in the wonderland of abstract algebra; but they serve. The realization that we are after a reasonable answer rather than a "one and only" right answer is an important step toward sophistication in the practical world of statistical inference.

Probabilities recorded in three or four decimal places are usually illusions of precision. Calculation to an accuracy of three or four significant digits can often lead to some pretty frail conclusions. Properly understood, though, our numerical computations give us a lot more information in the world of things than do descriptive adjectives.

In the seven chapters of Part I we have tried to pave the way to statistical inference by dealing with sampling theory, probability, frequency distributions, probability distributions, and mathematical models (binomial expansion, normal distribution, Tchebycheff's theorem).

The four chapters of Part II are concerned with statistical *estimation.* We shall be doing some statistically based guessing at population proportions from sample proportions, and at population means from sample means. We shall produce a numerical expression for the probability of being right or wrong (the calculated risk).

Clear organization of information, appropriate use of mathematical models, and realistic manipulation of arithmetic computations should increase your effectiveness in pursuing these goals.

Let's get to work.

• II •

STATISTICAL CONCEPTS FOR ESTIMATION

8

Distribution of Sample Proportions

Finally, and most important, the giver of the answer, call it Nature, is impersonal, impartial, and indifferent. She does not give opinions, or make judgments; she cannot be wheedled, bullied, or fooled; she does not get angry or disappointed; she does not praise or blame; she does not remember past failures or hold grudges; with her one always gets a fresh start, this time is the one that counts.

JOHN HOLT
Freedom and Beyond (1972)

OBJECTIVES

Given n (*the sample size*) *and* π (*the proportion of the population having a specified characteristic*), *to state the probability that the proportion in a random sample will fall in a specified interval of proportions.*

SYMBOLS AND FORMULAS FIRST USED IN THIS CHAPTER:

μ_p or $\bar{p}$	symbols for the mean of the proportions in different samples
$\bar{p} = \pi$	the mean of all sample proportions will be the same as the population proportion
$\sigma_p = \sqrt{\frac{\bar{p}(1-\bar{p})}{n}}$ or $\sigma_p = \sqrt{\frac{\pi(1-\pi)}{n}}$	the standard deviation of the proportions of all possible samples the same size
$z\sigma_p$	the distance of p from $\bar{p}$ (here again we can use $\bar{p}$ or π, because they have the same value)

8–1 EXERCISES

1. Give a rule of thumb for allowing the use of the normal curve as an approximation to the binomial.

2. In considering the distribution of sample proportions, what does $\bar{p}$ (p bar) stand for? What is the formula for the standard deviation of the sample proportions? What can you substitute for $\bar{p}$? Why?

3. A United States senator stated that 75% of our senior citizens (those 65 years of age or over) depend on Social Security alone for their subsistence. If this is so, what is the probability that in a random sample of 800 senior citizens, between 0.70 and 0.80 will be totally dependent on Social Security? (Sample was taken randomly from every state and Washington. D.C.—in each case, one senior citizen for about every 250,000 people.)

4. The U.S. Census Bureau reported late in 1970 that 30% of all householders own two cars. If 100 householders in the United States were selected randomly, what would be the probability that the sample proportion of householders owning two cars would be over 0.40? Use Table IV(b) in the appendix on page 237.

5. A white clover seed package is claimed to contain only 0.05 weed seeds. In any random sample, would you expect to get exactly $p = 0.05$? Try a sample of 10,000 grass seeds. (Get someone else to count them.) What is the probability you will get a p between 0.045 and 0.055? Give one reason why the large sample size helps to make this probability high?

6. Studies at a ski resort show that 2% of the skiers break a leg during the month of January. On this basis, what is the probability that 8 out of 300 weekend guests will break a leg? (The experience of the skier seems to have very little effect on this probability.)

7. Union cards were passed out to 2000 workers. The cards were returned, showing 1760 signed and 240 not signed; this is 88%. Their cards were then shuffled and a sample of 200 was drawn, showing 138 to be signed. Why would you conclude that random shuffling was not achieved?

8. An insurance company calculates that the probability of a man 40 years old living to be 60 is 0.70. If the company then insures 1000 men 40 years old, what is the probability, according to their calculations, that more than 20% will not live to 60?

9. Of the 200 million population in the United States in 1970, 1 million were said by a sociologist to be alcoholics. Calculate the individual probability based on chance alone. (Neolithic man used fermented berries; the ancient Assyrians sucked opium lozenges.)

10. Isopropyl Rubbing Alcohol is a 70% aqueous solution. Fifty bottles so labeled are selected randomly. What is the probability that the proportion of aqueous solution in the sample will be above 71%? No! Don't start. The type of problem we have been dealing with tells us the proportion of items in a population that do or do not have a certain attribute or characteristic. This problem states no such thing: it tells us the proportion of something in one item. Quite a different matter, requiring a different question.

    There will be another problem of this type later in the exercise. The proportion will be the proportion of something in a member of the population or sample, not a proportion of the members in the sample that have a given attribute. Watch out for it.

11. In the spring "Grapefruit League" a professional ballplayer guarantees that he will hit safely 0.325 of the time in official times at bat in the coming regular season. If he is right, what is the probability that he will hit over 0.390 in the first 50 times at bat in the regular season? If hitting is just a matter of luck, your prediction will be fairly good. If hitting in the long run is not luck, would your prediction be affected? Why, or why not?

12. A centerless grinding machine in a manufacturing plant turns out $\frac{1}{4}$-inch drill stock with 0.03 of it undersize. What is the probability that the next piece will be undersize? What is the probability that as large a proportion as 0.02 or more will be undersize in the next sample of 100?

13. In the type of problem we are presently dealing with (the distribution of sample proportions), it makes no difference *what* we are dealing with—peas or cannonballs. The standard deviation of the sample proportions is dependent only upon π and the sample size. If $\pi = 0.50$ and $n = 100$, what will σ_p be? If 0.50 of a population of peas is wrinkled, what is the probability that more than 0.55 in a sample of 100 will be wrinkled? If 0.50 of a population of cannonballs has no fuses, what is the probability that more than 0.55 will be without fuses in a sample of 100? Suppose you flip a half dollar 100 times. What is the probability that more than 0.55 of your tries will produce heads?

14. The chemical composition of the atmosphere is highly uniform for about 75 miles (120 kilometers) up. The largest part is nitrogen, about 78%.

What is the probability that the proportion of nitrogen in 100 randomly selected flasks will be above 80%?

15. If it is really true that 45% of the voters endorse Jones for governor of the state, what is the probability that a random sample of 400 voters will show 50% or more voting for Jones?

16. A floor inspector on a carpet tack machine is instructed to take a random sample of 100 tacks. If the process is putting out 2% defective tacks, the manufacturer wants to be 90% sure the process will be stopped. On what percent defective should the inspector stop the machine?

17. Dr. C. F. Westoff of Princeton University's Office of Population Research reported, in an analysis of the 1965 National Fertility Study, that 22% of all wedlock births were unwanted by at least one spouse. If the doctor's claim is correct, and if it is possible to get a random sample of 1000 births, what would be a reasonable probability that more than 0.24 of the births in the sample would be "unwanted"? The assumption is that this is a chance relationship. We could probably show, though, that the number of unwanted births is directly related to the number of previous children and inversely related to the economic and education level. Other relationships are involved, too, but at the present time we are concerned with what to expect if the relationship is just chance.

8–2 DISTRIBUTION OF SAMPLE PROPORTIONS

In samples, the proportions of members that have a selected characteristic will vary, though the samples continue to be taken from the very same population. A sample is simply not a reduced photograph of a population. This is a fundamental concept underlying statistical inference: draw samples from a population and your results will vary just by chance. These varying results may not indicate at all that the samples are coming from different populations or that somehow the sample source is changing. Then again, the results may vary enough to indicate that the samples do come from different or changing populations. In this latter case, the differences are said to be significant. It all depends on where you draw the line.

Red Fescue Grass Seeds

Red fescue is a variety of grass whose seeds are commonly mixed with other varieties of lawn seed. It is stated on some packages that 85% of the red fescue seeds will germinate.

Choose 100 red fescue grass seeds at random. (Rather difficult to find them in a mixed bag.) What proportion would you expect to germinate? You can see that if the claim is right, the sample proportions will clump around 0.85. Some will be a little farther away, and a very few will be extreme proportions (even 0.00 and 1.00 are theoretically possible—no red fescue seeds germinating, all red fescue seeds germinating).

Now the business of having most of the sample proportions cluster around the population proportion and, as we move away from π on either side, having the frequency of proportions fall off rapidly at first, then more and more slowly as we get farther and farther from the mean, suggests a normal distribution.

And that is just what happens. The distribution of sample proportions from *any* population approaches a normal distribution, particularly as the sample size increases. This is mathematically true. From amidst the apparent irregularities of nature comes floating the miraculous vision of a pattern, the curve of normal distribution, which you might have anticipated, depending, of course, on what you have been thinking about.

To approximate a normal distribution, you'd better be sure that $n\pi$ and $n(1 - \pi)$ are each greater than 5.

8–3 MEAN OF SAMPLE PROPORTIONS

The mean of sample proportions, μ_p or $\bar{p}$(p bar), is theoretically the same as the population proportion, π. If you think about it, you may arrive at the same truth intuitively on your own. As you keep drawing samples of grass seeds and keep adding their proportions and taking the mean of those proportions, you would indeed expect that mean to be coming closer and closer to $\pi = 0.85$. This can be shown mathematically to be true.

8–4 STANDARD DEVIATION OF SAMPLE PROPORTIONS

The mathematics will even show you that if you use the proportions of all the samples from a population, the standard deviation of these sample proportions will equal the square root of the fraction formed by the product of the population proportion and one minus that proportion divided by the sample size. Or simply:

$$\sigma_p = \sqrt{\frac{\pi(1 - \pi)}{n}}$$

This, of course, is derived from $\sigma_{np} = \sqrt{n\pi(1 - \pi)}$, the standard deviation for the *number* having a certain characteristic.

So now we can indicate the spread of the sample proportions of germinating red fescue seeds in samples of 100 by using the standard deviation of these proportions:

$$\sigma_p = \sqrt{\frac{\pi(1 - \pi)}{n}}$$

$$= \sqrt{\frac{(0.850)(0.150)}{100}}$$

$$= \sqrt{\frac{0.1275}{100}}$$

$$= \sqrt{0.001275}$$

$$= \sqrt{0.001280}$$

Thus

$$= 0.03578 = 0.0358$$

rounding off to three significant digits. (The zero before the 3 merely locates the decimal point; it is not a significant digit.) Here π equals the proportion

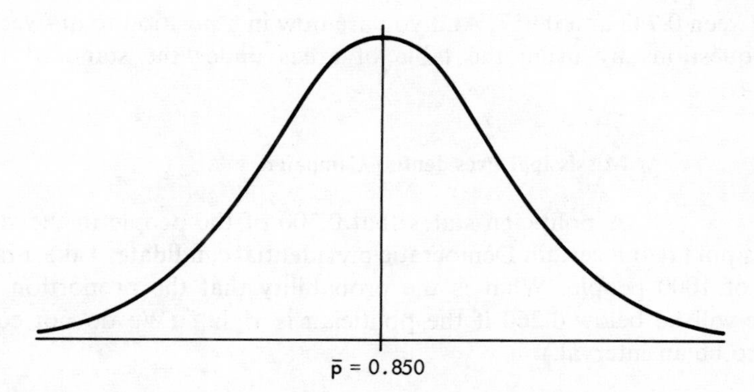

Distribution of sample proportions ($n = 100$, $\pi = 0.850$, $\sigma_p = 0.0358$)

of red fescue seeds that will germinate. We could have as reasonably based the problem on the proportion of seed that will *not* germinate.

As you know, for normal distributions, almost all the proportions will fall within three standard deviations of the mean of the proportions.

$$z\sigma_p = (3)(0.0358) \qquad = 0.1074 = 0.107$$

$$\bar{p} \pm z\sigma_p = 0.850 \pm 0.107 = 0.743 \text{ to } 0.957$$

We show the sample proportions as distributed continuously; we do not consider 0.743 to be an interval because we'd have to chop it down too fine (0.7425 and 0.7435).

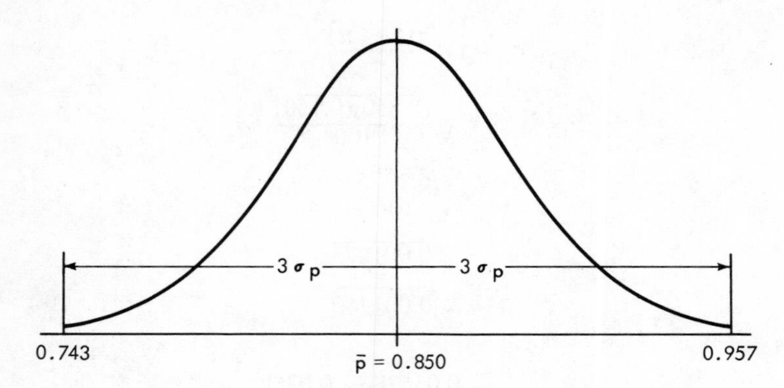

Distribution of sample proportions ($n = 100$, $\pi = 0.850$, $\sigma_p = 0.0358$)

Right off you get a good idea of the amount of variation among the sample proportions: more than 99% of the time sample proportions will fall between 0.743 and 0.957. And you are now in a position to answer many other questions by using the table of areas under the standard normal curve.

Mississippi Presidential Campaign

A politician states that 0.300 of the people in the state of Mississippi favor a certain Democratic presidential candidate. Take a random check of 1000 people. What is the probability that the proportion in the sample will be below 0.260 if the politician is right? (We do not consider 0.260 to be an interval.)

$$\pi = 0.300$$

$$\sigma_p = \sqrt{\frac{\pi(1-\pi)}{n}} = \sqrt{\frac{(0.300)(0.700)}{1000}}$$

$$= \sqrt{\frac{0.210}{1000}} = \sqrt{0.000210}$$

$$= 0.01449 = 0.0145$$

$$z = \frac{p - \bar{p}}{\sigma_p} = \frac{0.260 - 0.300}{0.0145}$$

$$= \frac{-0.0400}{0.0145} = -2.758 = -2.76$$

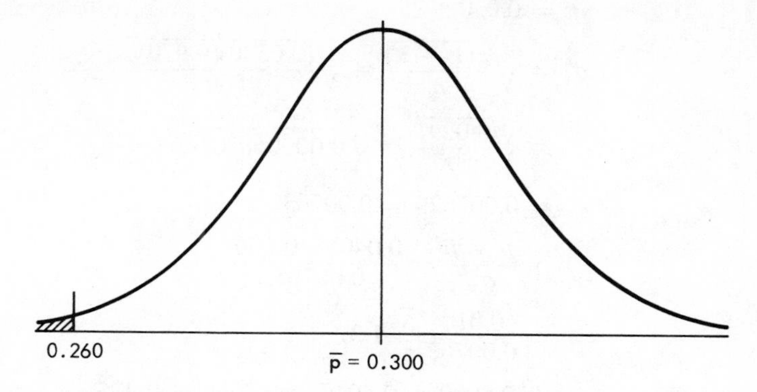

Distribution of sample proportions ($n = 1000$, $\pi = 0.300$, $\sigma_p = 0.0145$)

The area below 0.260($z = -2.76$) may readily be found in Table IV(b) to be 0.003 (see page 237 in the appendix). If the proportion in your sample is below 0.260, why should you be suspicious of the politician's statement?

Rejected Insurance Applications

A midwestern insurance company packs away all of its pink slips (rejected applications) in bundles of 500. Past experience shows that 3% of these rejected applications get to be erroneously filed. What is the probability that a randomly chosen bundle will show 4% or under to be erroneously filed?

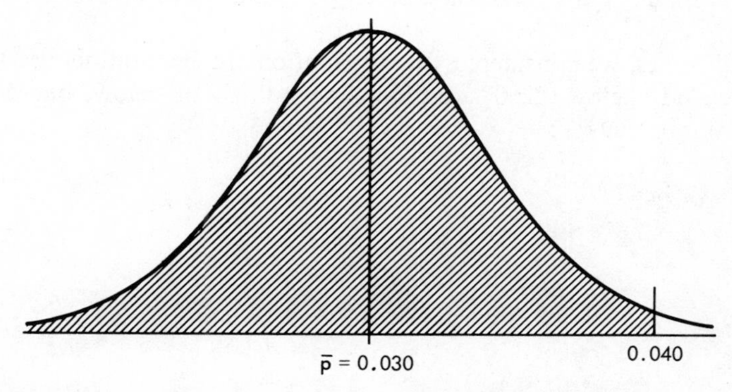

Distribution of sample proportions of rejected applications ($n = 500$, $\sigma_p = 0.00763$)

$$\pi = 0.030$$

$$\sigma_p = \sqrt{\frac{\pi(1-\pi)}{n}} = \sqrt{\frac{(0.0300)(0.970)}{500}}$$

$$= \sqrt{\frac{0.0291}{500}} = \sqrt{0.00005820}$$

$$= 0.007629 = 0.00763$$

$$z = \frac{p - \bar{p}}{\sigma_p} = \frac{0.040 - 0.030}{0.00763}$$

$$= \frac{0.010}{0.00763} = 1.31$$

$$P_{(4\% \text{ or under})} = 0.5000 + 0.4049 = 0.9049 = 0.905$$

8–5 SUMMARY

There is only one population proportion for a given characteristic, but if you take *samples* from that population, you may get as many different sample proportions as you have samples. Furthermore, as the number of samples increases, the mean of all these sample proportions will get closer and closer to the population proportion; we use $\bar{p} = \pi$.

As the sample size increases, the standard deviation of the proportions gets to be less and less, as shown by the formula (where the increasing denominator reduces the value of the fraction).

$$\sigma_p = \sqrt{\frac{\pi(1-\pi)}{n}}$$

Practically the only restrictive stipulation is that both $n\pi$ and $n(1 - \pi)$ be greater than 5. Otherwise you'd better go to a table of binomial probabilities.

In all cases, we considered the distributions to be continuous; that is, when we said "below 0.260" we didn't mean 0.259 or below, but *anything* below, even 0.25999. . . .

• 9 •

Distribution of Sample Means

What is man in nature? Nothing in relation to the infinite, everything in relation to nothing, a mean between nothing and everything.

BLAISE PASCAL (1623–1662)
Pensées, No. 72

OBJECTIVES

Given n (*the sample size*),
μ (*the mean of the population measurements*), *and*
σ (*the standard deviation of these population measurements*),
to state the probability that the mean of a random sample will fall in a specified interval of measurements.

SYMBOLS AND FORMULAS FIRST USED IN THIS CHAPTER:

$\mu_{\bar{x}} = \mu_x$	the mean of all the sample means is equal to the mean of the population
μ_x or μ	we often simply use the Greek *mu* for the mean of the population
$\sigma_{\bar{x}} = \dfrac{\sigma}{\sqrt{n}}$	the standard deviation of the sample means is equal to the standard deviation of the population divided by the square root of the sample size

9-1 EXERCISES

(Draw a graph every time you can, and label it clearly. This applies for the duration of this course and is designed to help you particularly, not me.)

1. If you take all possible samples from a given population, what can you say about the shape of the distribution of the means of these samples?

2. State a rule of thumb for determining whether the sample is large enough for us to use the normal curve as an approximation of the distribution of sample means.

3. If the sample is large relative to the population, you would have to use a correction factor in calculating the standard deviation. What rule of thumb would you apply here to determine when you can omit the correction factor? (The correction factor, $\sqrt{N - n/(N - 1)}$, is a computational device. We have so far to go that I shall not hold you up with problems that need this correction factor.)

4. How can you find the standard deviation of the sample means, given the population standard deviation?

5. The average weight of a bale of wastepaper in a certain company is supposed to be 200.0 pounds with a standard deviation of 15.0 pounds. Accepting this supposition as true, what would be the probability that a random sample of 36 would have a mean weight between 200 and 195 pounds? In this and most of the problems that follow concerning the distribution of sample means, consider these distributions to be continuous. Use the interval only if the probability of some particular mean is in question; don't use the interval if the probability in question is that of a mean more than or less than some value, or between two values.

6. A soft-drink machine in a student lounge is supposed to deliver exactly 7.00 ounces per cup. After extensive weighings, a student estimated the population standard deviation to be 0.53 ounces. Using this information, what would be the probability of getting a sample mean of between 6.98 and 7.02 in a random sample of 49?

7. For a certain variety of peas, the average pod length is 10.0 centimeters with a standard deviation of 1.3 centimeters. What is the probability that in a sample of 30 the average pod length will be over 10.6 centimeters?

8. A certain drug lowers human systolic blood pressure an average of 12 millimeters with a standard deviation of 5 millimeters. How likely is it that 64 patients will find that the drug lowers their systolic blood pressure an average of 10 millimeters (between 9.5 and 10.5)?

9. Of all registered voters in one congressional district, 64% are Republicans. What is the probability that a random sample of 100 of the voters in this district will be less than 60% Republican? (These are proportions, not means.)

10. The mean height of a population of grass blades is 7.5 centimeters. What is the probability that a random sample of 64 will have a mean between 7.1 and 7.9 centimeters?

11. If the average cost of a college textbook is $10.00 with a standard deviation of $0.50, at least what proportion of the books would you expect to find in the $9.00 to $11.00 interval? Watch out; it doesn't say the distribution is normal.

12. It is estimated that the average sales price of homes in Michigan in 1971 was $26,500. Assuming a normal distribution, in what price interval would you expect to find about 95% of the homes? Too bad; I would have been interested.

13. Suppose that on a true-or-false test with 10 questions, a student flips a coin on each question (heads, true; tails, false). What is the probability that he will pass—0.60 or more? Suppose there were 100 questions?

14. A company manufactures old-style BX cable averaging 47.1 turns per foot in the metal stripping of the outer covering with a standard deviation of 1.3 turns. How many turns would you expect to find in a sample of 100 randomly selected different feet of cable? Better use $z = 3$ if you want to be sure. This will have to be an interval estimate. Are you distinguishing between population and sample symbols?

15. A machine fills 16-ounce packages of baking soda. The resulting weights vary. Assuming a normal distribution with a standard deviation of 0.20 ounces, at what weight should the machine be set to risk underfilling only 5% of the time if the machine can be set accurately to the nearest hundredth of an ounce?

16. Certain steel-belted radial tires are supposed to last (?) 55,000 miles with a standard deviation of 5000 miles. What is the probability that the mean of a sample of 100 will be less than 49,000 miles? How about the probability of a single tire lasting less than 49,000 miles?

9–2 DISTRIBUTION OF SAMPLE MEANS

The mean of the measurements in one sample may vary from the mean of the measurements in another sample even though the samples come from the same population. This, again, is that important concept of chance variability. The means of samples the same size will vary—vary from each other and vary from the population mean just by chance. We do not want these differences to suggest that the samples did come from different populations. Only when the differences are *significantly* different shall we say that they indicate different populations. The level of significance will state the probability of being wrong.

Furthermore, the distributions of sample means take on a predictable form which the population may not have had at all—far from it. If you draw one card at a time from a pack of ordinary playing cards, recording its face value (1 through 13) and replacing that card before the next drawing, in the long run you will get just about as many of each card (kings, queens . . . sevens . . . twos, ones) because there are four of each (no more, no less) in the pack. The probability distribution is rectangular, quite different from a normal distribution, the probabilities of the different card values being the same in each case!

$$P_{(\text{any one card value})} = \tfrac{1}{13} = 0.077$$

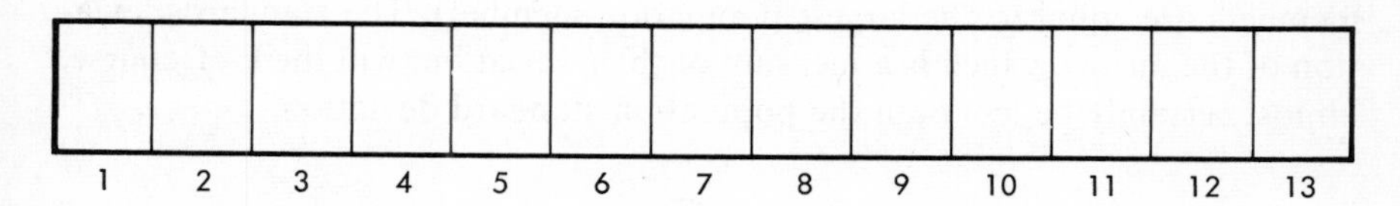

Probability distribution of card face values

We come upon different results when we plot sample statistics, say the *averages* of five cards at a time. The lowest average would have to be (1, 1, 1, 1, 2), or 1.2, and there are only four ways of getting this average (you are using all the 1's; there are four different 2's). There are also only four ways of getting 12.8 (13, 13, 13, 13, 12), the highest average, too. But the number of ways of getting other means in between increases as we go up from 1.2 to 7.0 and then decreases as we go down to 12.8. The most likely is 7.0, which can occur in a great many ways (10, 10, 10, 2, 3; 5, 6, 13, 6, 5; . . .).

Without pushing the mathematics of the situation, you will be right in suspecting this distribution of the means to grow more and more like the shape of a normal distribution as the sample size increases, though we started with a population distribution very much different from that of a normal distribution.

9–3 MEAN OF THE MEANS

If you take the mean of the means of all possible samples ($\mu_{\bar{x}}$), it will have the same value as the mean of the population (μ_x). Again, the theoretical mean of the sample means will equal the mean of the individual values in the population:

$$\mu_{\bar{x}} = \mu_x \quad \text{(mu sub } x \text{ bar = mu sub } x\text{)}$$

Think about this; as you take more and more samples, wouldn't you really expect the mean of these sample means to grow closer and closer to the true mean of the population?

9–4 STANDARD DEVIATION OF SAMPLE MEANS

In regard to the range of these means, you probably can see too that the sample mean will usually come from sample members that are larger and smaller than the mean (the mean of 3 and 7 is 5, which is larger than 3 and smaller than 7), so the smallest sample mean will be larger than the smallest member of the population (unless all the sample members are each equal to the smallest population member) and the largest sample mean will be less than the largest member of the population (unless all sample members are equal to the largest population member). The standard deviation of the means, which is a measure of their variation, will then, of course, almost certainly be less than the population standard deviation.

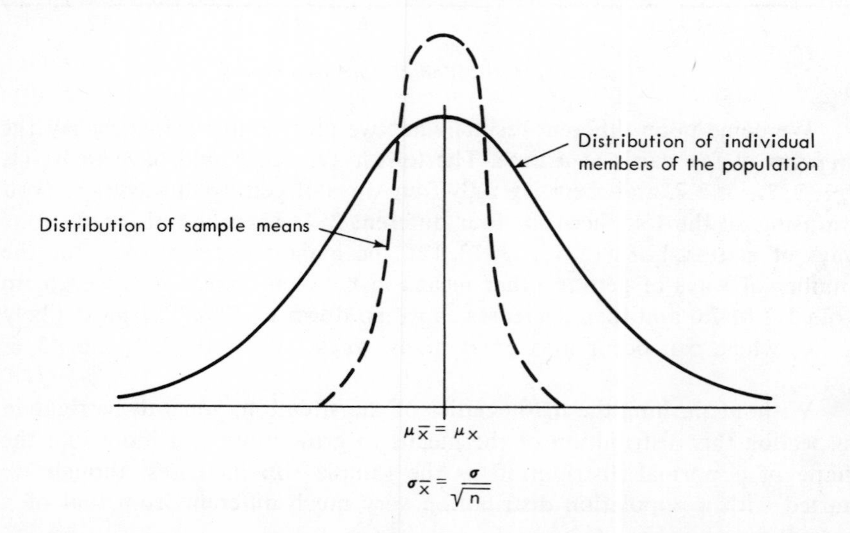

It turns out that you can get the standard deviation of the mean by dividing the population standard deviation by the square root of the sample size.

You can get the true standard deviation of the face values of individual cards, which is 3.74, in the following manner:

$$\sigma = \sqrt{\frac{\sum (x - \mu)^2}{N}}$$

$$= \sqrt{\frac{4(1-7)^2 + 4(2-7)^2 + \cdots + 4(12-7)^2 + 4(13-7)^2}{4(13)}}$$

$$= \sqrt{\frac{182}{13}} = 3.74$$

Then the standard deviation of samples of five will be

$$\sigma_{\bar{x}} = \frac{\sigma}{\sqrt{n}}$$

$$= \frac{3.74}{\sqrt{5}}$$

$$= \frac{3.74}{2.24}$$

$$= 1.67$$

The variation (the standard deviation) of the five-card sample means is not as much as the variation of individual card values, just as we had expected. Mathematical theory says $\sigma_{\bar{x}} = \sigma/\sqrt{n}$, so of course $\sigma_{\bar{x}}$ is probably smaller than σ.

Now it is important to reassure you that as the sample size increases, the distribution of the sample means tends more and more to take the form of a normal distribution, even though the population distribution may be far from normal. Notice also that the deviation from the normal is likely to be less when the sample is large. Try to keep the sample 30 or more.

A population distribution can have peculiar shapes like

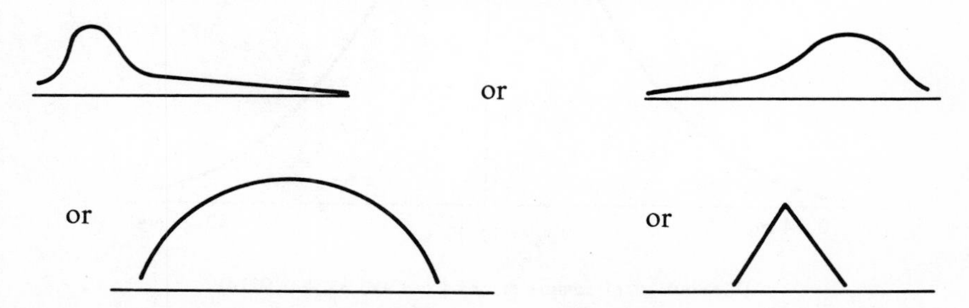

or even, as in card values,

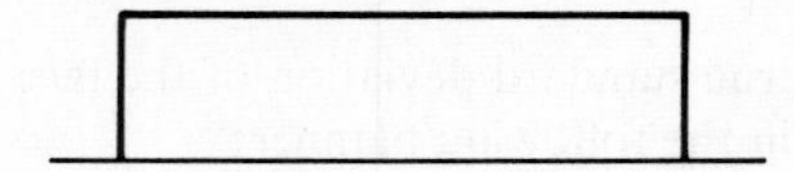

Nevertheless it can still be assumed that the distribution of the sample means is close to normal, that is, if the sample size is large ($n \geq 30$).

Vitamin C Tablets (ascorbic acid)

Two thousand vitamin C tablets show $\mu = 50.20$ milligrams with $\sigma = 0.15$ milligrams. Let's examine the distribution of the sample means, $n = 49$.

The sample is more than 30 (assume normality for distribution of sample means).

$$\mu_{\bar{x}} = \mu_x = 50.20 \text{ mg}$$

$$\sigma_{\bar{x}} = \frac{\sigma}{\sqrt{n}} = \frac{0.15}{\sqrt{49}} = \frac{0.15}{7} = 0.0214$$

If you want to guess at the total variation, you could use 3σ, which will give you 0.4987 of the area on each side of the mean.

$$\begin{aligned} \mu_{\bar{x}} \pm z\sigma_{\bar{x}} &= 50.20 \pm 3(0.0214) \\ &= 50.20 \pm 0.0642 \\ &= 50.14 \text{ to } 50.26 \text{ mg} \\ P_{\bar{x}(\text{within } 50.14 \text{ to } 50.26 \text{ mg})} &= 2(0.4987) = 0.997 \text{ or } 99^{+}\% \end{aligned}$$

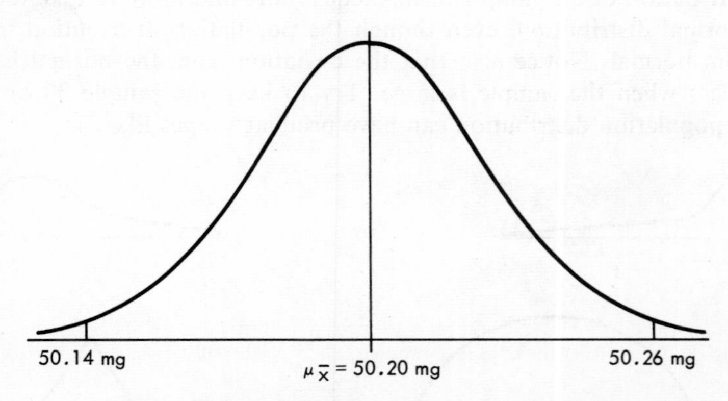

Distribution of sample means ($n = 49$, $\sigma_{\bar{x}} = 0.0214$)

The sample means will be within 50.14 and 50.26 milligrams 99.7% of the time. Pretty good security!

Pound of Butter (about 453.6 grams)

From long experience an inspector for the Weights and Measures Department expects a pound of butter to weigh about 1.05 pounds with a standard deviation of 0.093 pounds. How likely is it that a sample of 30 will show an average weight of less than 1 pound?

$$\mu_{\bar{x}} = 1.05 \text{ lb}$$

$$\sigma_{\bar{x}} = \frac{\sigma}{\sqrt{n}} = \frac{0.093}{5.48} = 0.0170$$

$$z = \frac{\bar{x} - \mu_{\bar{x}}}{\sigma \bar{x}} = \frac{1.00 - 1.05}{0.0170} = \frac{-0.050}{0.0170} = -2.94$$

Area below $z = -2.94$ is 0.0016

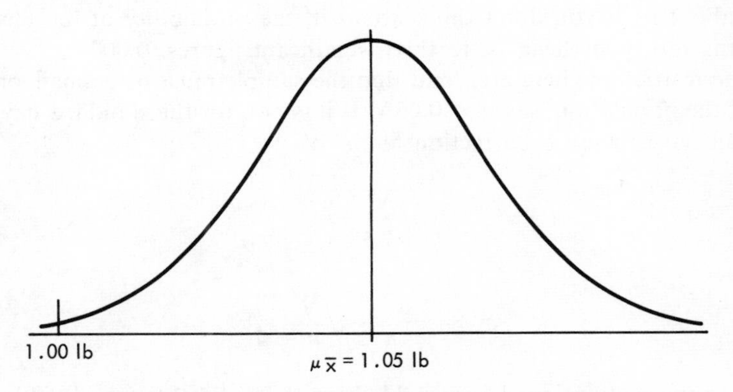

Distribution of sample means (n = 30, $\sigma_{\bar{x}}$ = 0.017 lb)

$$P_{(\bar{x} \text{ below } 1.00 \text{ lb})} = 0.0016$$

Safer than you thought, isn't it?

9–5 SUMMARY

Would you have believed it? Instead of getting more complex and complicated and hard to understand, statistical inference seems to be becoming simpler. No matter what the shape of population distributions, the distribution of sample means and sample proportions, under the right conditions, will approach the normal curve close enough for us to use it as an approximation.

In the case of proportions.

$$\bar{p} = \pi$$

$$\sigma_p = \sqrt{\frac{\pi(1 - \pi)}{n}}$$

In the case of means:

$$\mu_{\bar{x}} = \mu$$

$$\sigma_{\bar{x}} = \frac{\sigma}{\sqrt{n}}$$

And those are pretty simple tools, really.

You can see, if you think about it, that in a distribution of sample proportions most of them will fall around π. Then there will be a rapid falling off of frequencies, until the extremes stretch on and on. I said, if you think about it; if you don't think about it, the probability of the idea just dropping into your head is, to three significant figures, 0.000.

The restrictions here are, first, that the sample must be a small proportion of the population, say $n < 0.05N$. If it is not, for the standard deviation we shall have to use a correction factor of

$$\sqrt{\frac{N - n}{n - 1}}$$

that is,

$$\sigma_{\bar{x}} = \frac{\sigma}{\sqrt{n}} \cdot \sqrt{\frac{N - n}{n - 1}}$$

We keep the populations large in this book to avoid the considerable extra arithmetical exercise involved in using this correction factor.

To keep the distribution of the means close to normal we should take a sample of 30 or more. Also, we simplify matters somewhat by considering the distributions to be continuous, except when we are after the probability of a specific mean value.

Now for inferential estimation!

• 10 •

Estimation—Population Proportions

There are few things which we know, which are not capable of being reduc'd to a Mathematical Reasoning, and when they cannot, it's a sign our knowledge of them is very small and confus'd; and where a mathematical reasoning can be had, it's as great folly to make use of any other, as to grope for a thing in the dark, when you have a candle standing by you.

JOHN ARBUTHNOT (1667–1735)

OBJECTIVES

To state the probability that π *(the population proportion) lies in some specified confidence interval built around* p *(the sample proportion).*

SYMBOLS AND FORMULAS FIRST USED IN THIS CHAPTER:

$\hat{\sigma}_p = \sqrt{\frac{p(1-p)}{n}}$	the estimated standard deviation of the sample proportions, size n
$p \pm z\hat{\sigma}_p$	confidence interval estimate of population proportion

Here is a table of z's for commonly used areas beyond z standard deviations from the mean:

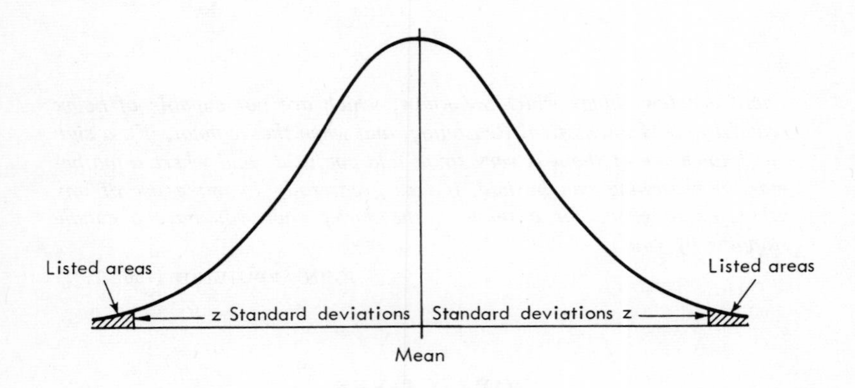

Areas	z
0.050	1.64
0.025	1.96
0.020	2.05
0.010	2.33
0.005	2.58

Even worth memorizing, as you will discover.

10–1 EXERCISES

1. What does a sample proportion tell you about the exact value of the proportion in the population from which the sample came?

2. Suppose you keep taking samples of the same size from a given population. What can you say about the distribution of sample proportions? Any restrictions?

3. What sample statistics do you use in setting up a confidence interval estimate? Do you need anything else?

4. What does a confidence interval estimate of a population parameter really say?

5. A random sample of 1000 childbirths in the United States taken in a sociological research project in 1965 showed 22% unwanted by at least one parent. Make a 99% confidence interval estimate of the national proportion of childbirths unwanted by at least one parent.

6. A random sample of fish species in a certain lake was attempted by the use of nets in different areas and at different depths. In a sample of 1000 it was found that 290 were members of the bass family. Make a 95% confidence interval estimate for the percentage of bass among fish in that lake.

7. In a random sample of 120 prescriptions written in the United States, 0.30 were for mood-altering pep pills or tranquilizers. Make a 98% confidence interval estimate of the proportion of prescriptions for pep pills or tranquilizers in the whole country.

8. A random sample of 150 wounded Marines in Vietnam as of November 1970 showed that 0.580 of them had to be hospitalized. Make a 95% confidence interval estimate of the proportion of wounded Marines in Vietnam who had to be hospitalized.

9. The mean of a population of test scores is 75 with a standard deviation of 10. What percent of scores will be between 55 and 95? Careful!

10. Under certain conditions, the probability that a salmon survives to maturity is 0.05. What is the probability that in a random sample of 100 newly hatched salmon 5, no more or less, will survive to maturity?

11. A marriage counselor claimed that in his community a sample of 100 women married to men 30 years or older showed that 75% objected to

pro-football doubleheaders on Sunday. Set up a 95% confidence interval estimate for the population proportion of such women who would object.

12. Under what conditions may these three distributions be considered approximately normal:
    (a) The binomial.
    (b) The proportions in samples of the same size from a given population.
    (c) The means of samples of the same size from a given population.

13. An insecticide is supposed to kill 70% of the insects sprayed (on or at?). How many could we expect with a 90% degree of confidence to be killed in a sample of 100? Notice that we ask how many, not what percent.

14. Records show that the suicide rate for 1951 to 1953 in Japan for males, ages 15 to 19, was 26.1 in 100,000. Set up a 95% confidence interval estimate for the number in 1 million.

10–2 ESTIMATION

We ask a sociologist to estimate the current proportion of ineligible welfare receivers in Massachusetts. He gets what he hopes to be a random sample of 800 welfare receivers in the state. The sample shows a proportion of 0.14 to be ineligible.

What does this sample tell us about the population? Doesn't it say only, at first, that in a randomly selected sample of 800 a proportion of 0.14 turned out to be ineligible? We know from our study of sampling distributions that the proportions in random samples of 800 will vary just by chance, conceivably going from as little as 0.00 to as much as 1.00. So picking 0.14 as the population proportion certainly wouldn't have a high probability of being exactly right. But knowing, as we do, that the distribution of sample proportions tends toward the normal, we could say that the probability of 0.14 being somewhere near the population mean is fairly good. That's *something* we can say from the sample about the population, but "somewhere near" and "fairly good" are rather loose descriptions.

Estimating by a single figure is called *point estimating*. It is often useful, but usually risky. Though sample counts (and therefore proportions) can be exact, it is highly improbable that the population proportion will be exactly the same as the sample proportion. It is highly improbable, too, that the population *mean* will be exactly the same as the sample mean. Moreover, in the case of means, the sample mean itself (being a measurement, not a count) cannot be exact. The statement that the mean weight of 100 cubic meters of cork is 191 kilograms cannot be exactly right even before we use it to estimate with.

We tell our sociologist that point estimates are risky. "Well, then," he says, "*about* 0.14." "What do you mean by 'about'?" He answers, "Oh, plus or minus 2%." Off the top of his head he has come up with an interval estimate: from 0.12 to 0.16 ineligible receivers.

We suggest to him that there may be statistical techniques that will enable us to make more realistic interval estimates. We say this because of our knowing that the distribution of sample proportions tends to be normal. Perhaps we can even come up with a numerical estimate of the probability of being wrong.

10–3 DISTRIBUTION OF SAMPLE PROPORTIONS

Success in any area—work, sports, marriage—is often, admittedly, due to good guessing. And there are some experienced people who become exceptionally good guessers and are much in demand—vice presidents, baseball managers, marriage counselors. But in many areas there are basic statistical concepts (confidence interval estimation, for instance), which help make us all much better guessers.

You already know that the probability distribution of all the proportions in samples from a given population will be approximately normal and will have a mean proportion and standard deviation that depend upon the population parameter $\bar{p}$ and the sample size n.

$$\bar{p} = \pi$$

$$\sigma_p = \sqrt{\frac{\bar{p}(1 - \bar{p})}{n}}$$

$$\text{or} \quad \sqrt{\frac{\pi(1 - \pi)}{n}}$$

because the true $\bar{p}$ will equal π.

Assume, for the purpose of discussion, that π actually does equal 0.14.

$$\begin{aligned}
\sigma_p &= \sqrt{\frac{\pi(1 - \pi)}{n}} \\
&= \sqrt{\frac{(0.14)(0.86)}{800}} \\
&= \sqrt{\frac{0.120}{800}} \\
&= \sqrt{0.000150} \\
&= 0.0122
\end{aligned}$$

In what interval would you expect to find 99% of the proportions if you go from $\bar{p}$ the same distance to the right and left? A 99% confidence interval estimate will leave 0.005 beyond $z\sigma_p$ on each side, so use $z = 2.58$. Remember?

$$\bar{p} \pm z\sigma_p$$
$$0.14 \pm 2.58(0.0122)$$
$$0.14 \pm 0.0315$$
$$0.11 \text{ to } 0.17$$

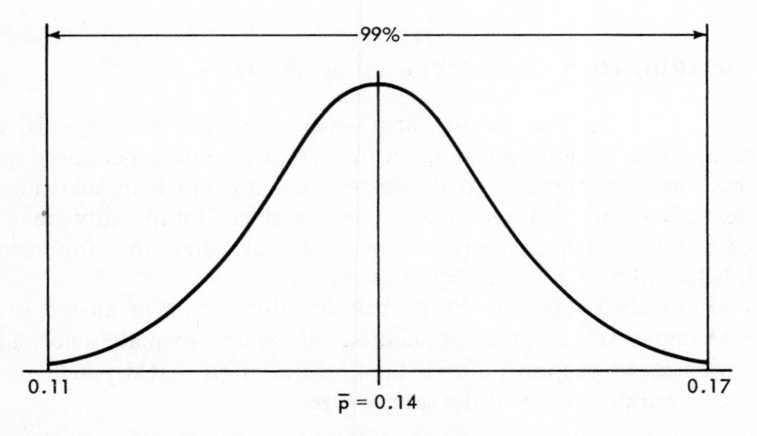

Distribution of sample proportions of ineligible welfare receivers ($n = 800$, $\sigma_p = 0.0122$)

To repeat: If $n = 800$ and $\pi = 0.14$, we would expect, on repeated samplings, that 99% of the sample proportions would fall between 0.11 and 0.17; only 1% of the proportions would fall outside this interval in the long run.

10-4 CONFIDENCE INTERVALS

Now take it slowly. If π really does equal 0.14 and if you take a sample of 800, the proportion can fall anywhere on the horizontal axis between 0.00 and 1.00 just by chance; but 99% of the time this sample proportion should fall somewhere in the interval 0.11 to 0.17. And when it does fall in this interval, say at $p = 0.13$, the previously calculated interval ($p \pm z\sigma_p$) now centered at $p = 0.13$ will include the true population proportion (that is, the true population proportion will probably—99% probably—lie in this interval).

Look carefully. Putting the center of the interval on either $p = 0.13$ or $p = 0.16$ shows the interval then including 0.14, the true population proportion. Whatever p you choose between 0.11 and 0.17 on which to place the center of the calculated interval, the interval will include $\pi(0.14)$. And 99% of the time p *will* be between 0.11 and 0.17. So you see that 99% of the time your interval will catch the population proportion.

But when a sample proportion falls outside the original 0.11 to 0.17 interval, say $p = 0.19$, then the sample interval estimate will not include the population proportion.

It might look like this, 0.14 lying outside the interval:

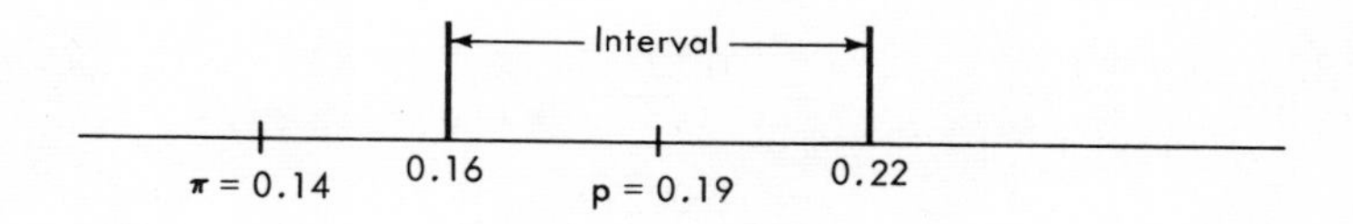

Because the interval is centered on random sample proportions, it will include π only 99% of the time. For 1% of the time, it will not catch π, and the interval will not work. The calculated risk of being wrong is 1%.

The difference between real estimation and what we have been doing, of course, is that we don't have π to estimate π with; we have to use p for π in estimating σ_p. Now this won't be just right, but it's very likely to be close to it, and it's all we've got. Put a hat on sigma to ensure remembering that it is an estimated value ($\hat{\sigma}_p$).

$$\hat{\sigma}_p = \sqrt{\frac{p(1-p)}{n}}$$

$$p \pm z\hat{\sigma}_p$$

Pure mathematics thrives on abstraction in a known world. Statistical inference is involved with concrete information obtained from variables in an uncertain world. If you don't have what you need, you have to find some reasonable substitute or go without.

Incidentally, the sociologist must have been one of those good guessers: he got 0.12 to 0.16 compared with our labored 0.11 to 0.17. But we went him one better by saying *how* confident we were: 99%. (Of course, we could have let him guess that, too.)

Marijuana Users

A random sample of 95 marijuana users showed that a proportion of 0.65 of them never became seriously involved or went on to some other drug. Show a 98% confidence interval estimate of the proportion of all marijuana users who don't become seriously involved or don't go on to some other drug.

You can look up the appropriate z in the standard normal distribution table in the appendix (page 237) or in the abbreviated table at the beginning of this chapter. Much better, and easier for you, is to memorize the abbreviated table at the beginning of this chapter. The z for 0.01 on one side of the mean is 2.33.

$$\begin{aligned}\hat{\sigma}_p &= \sqrt{\frac{p(1-p)}{n}} \\ &= \sqrt{\frac{(0.650)(0.350)}{95}} \\ &= \sqrt{\frac{0.2275}{95}} \\ &= \sqrt{0.00239} \\ &= 0.0489 \\ z\hat{\sigma}_p &= 2.33(0.0489) \\ &= 0.114 \\ &= 0.11\end{aligned}$$

Our 98% confidence interval could be

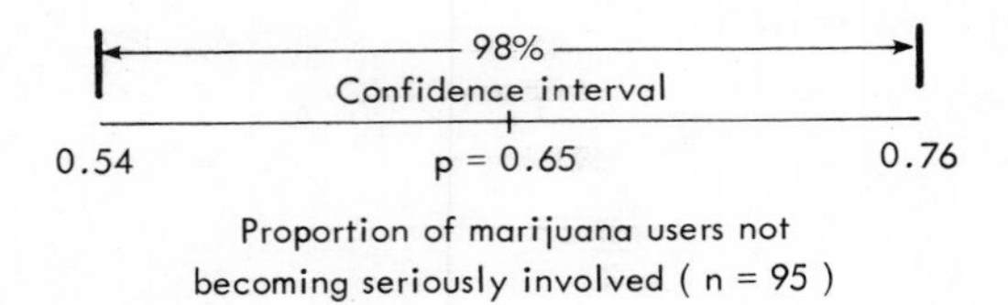

When you reduce the percent of confidence, you have to reduce z, and that means you have a smaller interval. As you make the interval wider, increasing z, your confidence goes up. The interval $p \pm 3\sigma_p$ will practically always catch π; the interval $p \pm 0\sigma_p$ becomes a point estimate. A point estimate certainly pins down the estimate closely, but it is almost never right. An interval estimate may be true a high percent of the time, but it doesn't pin down the estimate so closely. As in daily life, you can be a lot surer if you're not too fussy about how close you come to being right. The wider the interval, the greater your confidence.

A 90% confidence interval for the proportion of marijuana users who go on taking drugs would be

$$\begin{aligned}p \pm z\sigma_p &= 0.65 \pm 1.64(0.0489) \\ &= 0.65 \pm 0.080 \\ &= 0.57 \text{ to } 0.73\end{aligned}$$

This interval is a tighter guess, but we have less confidence in it. A small target is less likely to be hit. (William Tell and the apple on his son's head.) Look up *stochastic.*

10–5 SUMMARY

The distribution of sample proportions from a given population will have an approximately normal probability distribution. Try to keep np and $n(1 - p)$ both larger than 5 to ensure a satisfactory approximation.

The mean of the sample proportions will equal the population proportion.

$$\bar{p} = \pi$$

Estimate the standard deviation of the sample proportions by using p for π.

$$\hat{\sigma}_p = \sqrt{\frac{p(1 - p)}{n}}$$

The improbability of the population proportion being exactly the same as the sample proportion urges using an interval estimate rather than a point estimate. Obtain this interval by using the sample proportion, the sample size, and by assuming the distribution of the sample proportions to be normal.

Specifying a confidence of say, 90%, makes the calculated risk of being wrong 10%. As the interval *narrows*, the danger of turning down the true π increases: this is called a Type I error. If the interval *widens*, the danger of accepting a false π increases: this is called a Type II error. More on this later.

Remember that the area beyond one of the two confidence interval limits is given by these z's:

Area Beyond One Limit	z
0.050	1.64
0.025	1.96
0.020	2.05
0.010	2.33
0.005	2.58

In summarizing, we should admit that in order to compute, we have to use point estimates. If the grades in your statistics class are given by letters A, B, C, D, F (meaning 100–90, 89–80, 79–70, 69–60, 59 or below), you can't compute the class average. Given point estimates of the class grades (93, 42, 86 . . .), then the class mean and standard deviation, and so on, can be computed. But often this point estimating sacrifices real confidence for the illusory security of the results of numerical calculations.

It is in the same way that statistics for insurance rates, defectives in an industrial process, and estimates of gas mileage for airplane travel, rely on point estimates to allow for calculations. But when estimation is the ultimate objective, when it is to be used as illumination rather than for further calculation, then the interval estimate is often clearer and safer, and it enables us to come up with a numerical expression of the calculated risk of being wrong.

We have arrived, at last, in the complex land of statistical inference. Now this chapter and the ones that follow will generate more and more techniques for the decision-making process. When you face an unannounced type of problem, you have to decide what type it is (a most important step) and then select the appropriate technique for its solution. Rather than fishing around in your notes or the text, turn to the symbolic summary of technical procedures on the last pages of our appendix. Turn to it right now. Find the section for estimating population proportions from sample proportions. Use it as you do the exercises. The more you use it, the better you'll like it.

• 11 •

Estimation—Population Means

The more progress physical sciences make, the more they tend to enter the domain of mathematics, which is a kind of centre to which they all converge. We may even judge of the degree of perfection to which a science has arrived by the facility with which it may be submitted to calculation.

A. QUETELET (1796–1874)
[Belgian astronomer—perhaps the first to make use of the normal distribution in social studies]

OBJECTIVES

To state the probability that μ (the population mean) will fall in some specified confidence interval built around $\bar{x}$ (the sample mean).

SYMBOLS AND FORMULAS FIRST USED IN THIS CHAPTER:

$\bar{x} \pm z\hat{\sigma}_{\bar{x}}$	our confidence interval estimate of the population mean
$s = \sqrt{\frac{(x - \bar{x})^2}{n - 1}}$	the sample standard deviation used in place of the population standard deviation
$\hat{\sigma}_{\bar{x}} = \frac{s}{\sqrt{n}}$	our estimate of the standard deviation of the means
t	the number of standard deviations we use on small samples instead of z

11–1 EXERCISES

1. What sample statistics will you need in order to estimate the population mean? What other information do you need?

2. Can you justify using s for σ in computing $\sigma_{\bar{x}}$? How?

3. In checking his gas mileage, a driver got a mean of 20 miles per gallon for 50 gallons. What is the sample size? What was the total mileage (Σx)? What is the population? How would the standard deviation of the population compare with the standard deviation of the means of samples of 50 gallons?

4. A biologist took a random sample of 200 pumpkinseeds (fresh-water sunfishes) and found the mean length to be 150 millimeters (50 mm would be about 2 inches) with a standard deviation of 20 millimeters. Make a 95% confidence interval estimate of the mean length of pumpkinseeds.

5. Stewed tomatoes are often sold in small cans marked "net wt. $8\frac{1}{4}$ oz." A random sample of 144 cans of this product was checked for drained weight (weight of contents alone). The results turned out to be $\bar{x} = 8.50$ ounces, $s = 0.20$ ounces. Show a 99% confidence interval for an estimate of the population mean.

6. A state highway department tested the durability of test strips painted across a heavily traveled highway in 36 different places. The sample showed a mean life of 163,085 cars with a standard deviation of 13,914 cars. Find a 0.95 confidence interval for an estimate of the average number of cars it will take to wear off this paint. You might reasonably cut down the number of significant digits.

7. The volume of cells in the blood from 25 women between 18 and 23 years of age taken at random in the city of Chicago averaged 41.0 cubic centimeters per 100 cc. Make a 98% confidence interval estimate, using the t distribution, of the average volume of cells per 100 cc in the blood of women this age. To what women, and where, would you say this estimate applies?

8. An efficiency expert in a machine tool plant wants to estimate the average time it takes a machinist to shape a small gear. Timing a random sample of 50, the machinist's production efforts showed $\bar{x} = 18$ minutes with $s = 2.3$. Make a 95% confidence interval estimate of the population mean.

9. In order to test the durability of a new type of suitcase in air transportation, a random sample of 100 pieces of this new product were tested. They averaged 423 loadings before showing serious defects. The sample standard deviation was 28 loadings. Make a 95% confidence interval estimate of the mean number of loadings.

10. A small store averaged \$37.23 per week for janitorial service ($s =$ \$10.30) over a period of 36 weeks. Construct a 99% confidence interval for the store's estimated weekly cost of janitorial service.

11. To check the accuracy of his ledger, an accountant takes a random sample of 1500 entries from a total of about 110,000 entries. If more than 2% are in error, he checks all of the entries; if 2% or less are in error, he accepts the ledger as accurate. What is the probability that he will accept the ledger when actually 3% of the entries are in error?

12. Subway trains in one direction arrive at a certain station on the average of 15 minutes apart with a standard deviation of 3 minutes. What is the probability that just having missed one train, you will have to wait 20 minutes or more for another, assuming a normal distribution of the time between arrivals?

13. A friend gave me a small bag of English walnuts, promising that 4 out of 5 would grow. There were 14 walnuts in the bag; what is the probability that just half of them would grow? What is the most probable number to grow? Assume $\pi = 0.80$ to be true.

14. The mean age of 100 men engaged in steeplejacking is 55.2 years. Make a 99% confidence interval estimate for the mean age of all men engaged in the occupation.

15. A random sample of 25 children, from 5 to 10 years old inclusive, in a small Delaware town was checked for the hours a week they watched TV. The mean came out 15 hours with a standard deviation of 2 hours. Make a 98% confidence interval estimate of the population mean. To what children does this estimate apply? Why?

16. If 1.02 pounds is the average weight of "one pound" of Gilt Cow butter, what is the probability that an inspector from the Weights and Measures Department, in selecting one package, will find it underweight? Past experience shows $\sigma = 0.01$ and the distribution to be approximately normal.

17. In the previous problem, suppose the inspector had taken a sample of 30. What is the probability that the sample mean will be underweight?

18. Suppose, in the Gilt Cow butter situation, the population mean actually was $\mu = 1.00$. What would be the probability of getting a sample mean ($n = 30$) underweight?

19. What sample average weight should the producer shoot for so that the likelihood of the inspector getting an underweight mean ($n = 30$) is 0.05?

20. Ten safety belts manufactured by Fabrics, Inc. showed an average breaking strength of 8000 pounds with a standard deviation of 100 pounds. Estimate a 95% confidence interval for the population's mean breaking strength.

11–2 ESTIMATION

Light Bulbs

A light bulb manufacturer took a random sample of 100 of his first 10,000 new-model 75-watt bulbs, getting a mean life of 2310 hours with a standard deviation of 215 hours.

What does the sample suggest as a population mean? What is the likelihood that the sample mean will be the same as the population mean? Very unlikely; you certainly know that. What is the likelihood that the sample mean will be near the population mean? Rather good that it will be somewhere near it. What is the probability that the sample mean will be quite a distance from the population mean? It could happen all right, but as you go farther away from the population mean, the sample mean becomes less and less likely.

These nonnumerical descriptions, such as "very likely," are not really satisfactory. From what we have learned about probability and the distribution of sample means, we should be able to make specific confidence interval estimates. Let's try for a 95% confidence of being right.

11–3 DISTRIBUTION OF SAMPLE MEANS

We spent Chapter 9 discussing this distribution of sample means. The sample means, as do the sample proportions, approach a normal distribution. (Galton again!) We assume a normal distribution as long as the sample is large, say $n \geq 30$, and we use $\sigma_{\bar{x}} = \sigma/\sqrt{n}$ without a correction factor if the sample is not too large in relation to the population, say $n < 0.05N$.

In our light bulb situation,

$$n = 100$$
$$n \geq 30$$
$$n < 0.05N,\ 30 < 0.05(10{,}000),\ 30 < 500$$

Remember that $\sigma_{\bar{x}} = \sigma/\sqrt{n}$ (sigma sub x bar equals sigma of the population divided by the square root of the sample size) and $\mu_{\bar{x}} = \mu$ (mu sub x bar equals mu, the mean of the means equals the population mean).

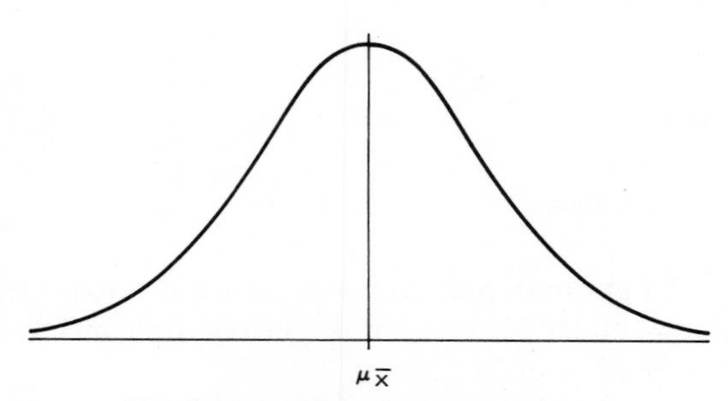

Distribution of sample means in hours

11–4 CONFIDENCE INTERVALS

We keep going over the same or similar concepts. If you are lost at first, hang in there; things are likely to clear up. And this approach, hopefully, will make you remember some of the important concepts of statistical inference long after the final exam.

We are again concerned with confidence interval estimates but this time it is for population means from sample means. You will need $\bar{x}$, n, and $\sigma_{\bar{x}}$. Go a certain distance to each side of the sample mean to form the confidence interval and you may catch the population mean in it. The size of this interval estimate is dependent upon the degree of confidence that you select as well as $\bar{x}$, n, and $\sigma_{\bar{x}}$.

A 95% confidence interval estimate for mean light bulb life:

$$\bar{x} \pm z\sigma_{\bar{x}} = 2310 \pm 1.96\frac{\sigma}{\sqrt{n}}$$

But we don't have σ!

So here we go again. What can we use for σ? Obviously, the sample standard deviation, which won't be just right but is likely to be nearly right. In fact, statisticians tell us that if in the formula for sample standard deviation we use $n - 1$ for n, it will be an even better estimate.

$$s = \sqrt{\frac{\Sigma(x - \bar{x})^2}{n - 1}}$$

The *sample* mean is $\bar{x}$. We'll use s for the sample standard deviation. You can see that as the sample standard deviation is probably less than the whole population standard deviation, dividing by $n - 1$ (the number of degrees of freedom) instead of by n will help by making it larger, thus improving its effectiveness in the stochastic (?) process. More on degrees of freedom in a few minutes.

Of course, if s is given, as is usually the case in this book, you won't have to worry about how to get it.

Use $\hat{\sigma}_{\bar{x}}$ (sigma sub $\bar{x}$ hat) to represent the estimated standard deviation of the sample means:

$$\hat{\sigma}_{\bar{x}} = \frac{s}{\sqrt{n}}$$

$$= \frac{215}{\sqrt{100}}$$

$$= 21.5 \text{ hours}$$

We take a 95% degree of confidence? That will give us 5% altogether, or $2\frac{1}{2}$% of the total area way off on each side of the sample mean, where we don't expect to catch the population mean. If the population mean actually lies way out there, we will be wrong, because we will be rejecting it. This is a Type I error whose total probability we allow to be 5%—$2\frac{1}{2}$% on each side.

$$\bar{x} \pm z\hat{\sigma}_{\bar{x}}$$

$$2310 \pm 1.96(21.5)$$

$$2310 \pm 42.14$$

$$(2268 \text{ to } 2352 \text{ hours})$$

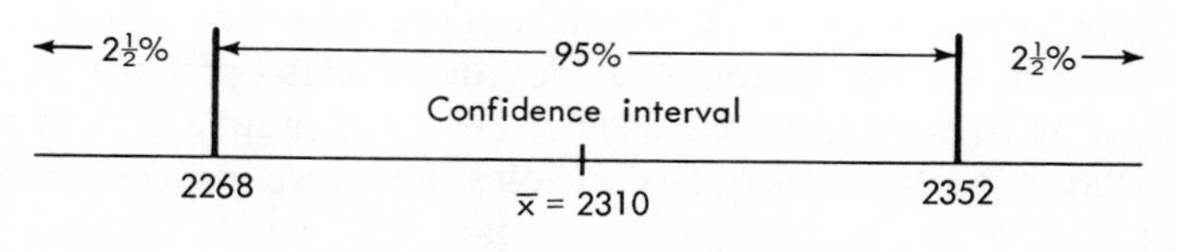

We estimate that the population mean is between 2268 and 2352 hours, inclusive. If it isn't, we will be wrong; and 5% of the time it won't be.

11-5 THE LAW OF LARGE NUMBERS

A spread of only 84 hours? Would you really expect the life of light bulbs to be that close? No, it surprised me at first, too. But what we've got here is not the variation of individual light bulbs; it is the variation of the means of samples of 100 light bulbs.

Assume that the real σ *is* equal to 215 hours.

$$\sigma_{\bar{x}} = \frac{\sigma}{\sqrt{n}}$$

You can readily see that as n increases, $\sigma_{\bar{x}}$ will get smaller. If the sample size is 10, the sample means will vary considerably; $\sigma_{\bar{x}}$ will be 68.0 hours. If the sample size is 100, the variation, as you know, is only 21.5 hours. If the sample size were 1,000,000, the variation of sample means would be almost imperceptible—depending of course on what you measure it with.

$$\sigma_{\bar{x}} = \frac{\sigma}{\sqrt{n}} = \frac{215}{\sqrt{1{,}000{,}000}} = \frac{215}{1000} = 0.215 \text{ hours}$$

So 99% of all the sample means fall within $\mu_{\bar{x}} \pm 2.58\sigma_{\bar{x}}$. Hence, $2310 - 2.58(0.215)$ and $2310 + 2.58(0.215)$, or $2310 - 0.555$ to $2310 + 0.555$, or 2309.445 to 2310.555—that close for the variation of sample means when the sample size is 1,000,000!

Taking a sample of 1,000,000 is not very practical. But I wanted you to see how increasing the sample size reduces the variability of the sample means. Though the life of individual light bulbs may vary considerably, the means of samples of 100 will vary only $\frac{1}{10}$ as much: $\sigma_{\bar{x}} = (\sigma/\sqrt{n})$. This is the statistical concept we used in our introduction to explain the reliability of cables on the George Washington Bridge. There may be considerable variation in the tensile strength of individual wires, but the means of samples of 26,464 will vary only $(1/\sqrt{26{,}464})$ as much. This is a statistical concept to remember.

When we use $\sigma_{\bar{x}} = (\sigma/\sqrt{n})$, we make the standard deviation of sample means dependent on the sample size, not dependent upon the population size. A question in manufacturing used to be: If we sample instead of doing 100% inspection, what proportion of each lot should we inspect? In industry, some said 20% of the population; some said 5%; but the commonest engineering rule was 10%. From a statistical point of view, however, this is like asking what proportion of the sugar bowl contents you should put in your coffee. It is the size of the sample that counts, not the size of the population; it is the teaspoon for your coffee, not the sugar bowl. A simple answer turned out to be: acceptance sampling of large lots should be based on actual sample size, not on their proportion of the population. As not infrequently happens in probability and statistics, the truth shows the unreliability of what we are accustomed to call "common sense," for it is the sample size that gives us our security, not what proportion of the population the sample is.

We have been citing the Law of Large Numbers. It says, in effect, what you already know—that the larger the sample size, the less the sample proportions and sample means will vary.

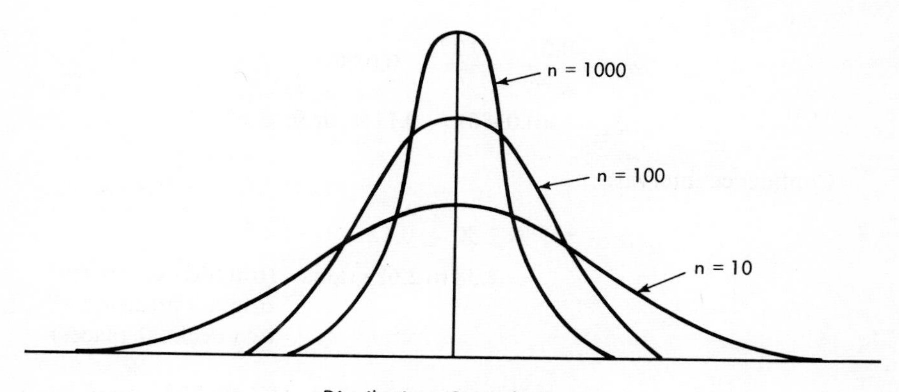

Distribution of sample means

It is a misinterpretation of the Law of Large Numbers when somebody cites "the law of averages" as showing that heads are due or that a hitter is due to hit. If you flip seven heads in a row, the probability of a head the next time is really still 0.500, but you erroneously think it less probable than that because you have already had so many heads that you expect tails to be showing up. If a 300 baseball batter strikes out seven times in a row, you know nothing about his getting a hit the next time up; you only know that in the long run, if he maintains that 300 average, he will have to get hits 3 out of 10 times at bat. But this prompts the fan to say that the batter is due. The Law of Large Numbers really tells us only about the long run, nothing about the next time.

So shall we strive for large samples? True that the larger the sample, the better the confidence interval estimate of the population proportion or the population mean. However, increasing the sample size much beyond a certain point usually does not have an important enough effect on the standard deviation to warrant the additional expenditure of time, labor, and money; we reach a point of diminishing returns.

Cold Tablets

We are interested in No-Drip cold tablets. $n = 49$, $\bar{x} = 2.50$ grains, $s = 0.27$ grains. (A grain is a pharmaceutical measure just under 65 milligrams.) To seek out a confidence interval we *could* use $z = 3.00$, which will give us practically a sure thing—3 chances in 1000 of being wrong. So, let's try it and go $3\sigma_{\bar{x}}$ on both sides of the sample mean.

$$\hat{\sigma}_{\bar{x}} = \frac{s}{\sqrt{n}}$$

$$\hat{\sigma}_{\bar{x}} = \frac{0.27}{\sqrt{49}} = \frac{0.27}{7} = 0.0386$$

$$z\hat{\sigma}_{\bar{x}} = 3(0.0386) = 0.1158 \text{ or } 0.12$$

Confidence interval:

$$\bar{x} \pm z\hat{\sigma}_{\bar{x}} = 2.50 \pm 0.12$$

$$= 2.38 \text{ to } 2.62 \text{ grains}$$ (rounded off to the original precision of two decimal places)

How would you characterize this interval? The 99.73% confidence interval? It is sometimes referred to as the 3-sigma interval.

Now suppose we had used $n = 100$, still with $s = 0.27$ grains.

$$\hat{\sigma}_{\bar{x}} = \frac{s}{\sqrt{n}}$$

$$= \frac{0.27}{\sqrt{100}} = \frac{0.27}{10} = 0.027$$

$$\bar{x} \pm z\hat{\sigma}_{\bar{x}} = 2.50 \pm 3(0.027)$$

$$= 2.50 \pm 0.081$$

$$= 2.42 \text{ to } 2.58$$

Whether the narrowing of the interval is important enough to warrant the increased cost of sampling will have to be determined by practical considerations.

Cold Cereal Consumption

What is the average cold cereal consumption of families in the United States?

Take a random sample of 1000 families and find the mean consumption per month to be 11.4 kilograms (about 400 oz or 25 lb) with $s = 3.3$ kilograms. Make a 98% confidence interval estimate of μ.

$$\bar{x} = 11.4$$

$$z\hat{\sigma}_{\bar{x}} = z\frac{s}{\sqrt{n}} = 2.33\left(\frac{3.3}{31.6}\right)$$

$$= 2.33(0.104) = 0.242 = 0.24$$

$$\bar{x} \pm z\hat{\sigma}_{\bar{x}} = 11.4 \pm 0.24 = 11.16 \text{ to } 11.64 \text{ kg}$$

(or possibly, to be on the safe side, 11.2 to 11.7 kg)

11–6 THE t DISTRIBUTION

When we replaced σ with the undoubtedly smaller s, we tried to compensate for it by using $n - 1$ in the denominator of the s formula. What statisticians call the number of **degrees of freedom** is, in this case, $n - 1$. Given $\bar{x}$ and n, you, or the Goddess of Chance, start picking values of the variable; after $n - 1$ choices you are done because the last value of x is predetermined by the given $\bar{x}$.

Suppose that $\bar{x} = 5$ and $n = 4$. The first three x's turn out to be 3, 4, and 6. But

$$\frac{3 + 4 + 6 + x}{4}$$

must be equal to 5. Then the last x must be 7.

$$\frac{3 + 4 + 6 + 7}{4} = 5$$

So $\bar{x}$ and $n - 1$ choices determined the nth x.

In the famous t distribution, the shape of the distribution is determined by the number of degrees of freedom. A distribution of sample means is close to normal when $n \geq 30$. Below $n = 30$, though still symmetrical, it begins flattening out more than the normal, making the use of z no longer reliable. Use t instead. The value of t will be determined by the number of degrees of freedom ($n - 1$) and, of course, the degree of confidence.

In 1908 William S. Gosset, an employee of the Guinness Brewing Company in Dublin, published a paper showing that when n is less than 30, though the normal distribution for the sample means cannot be used safely, this other distribution, dependent on the *number of degrees of freedom*, can be used. The Guinness Company had a rule forbidding publication of independent research; so Gosset signed his paper "Student." His discovery is often called *Student's t distribution.*

See Table IV(c) on page 238 in the appendix for our t table. Suppose you want the t for a 95% confidence interval with a sample of 15. Find the $t_{0.025}$ column, because 0.95 leaves half of 0.05 or 0.025 in each end area under the curve. Now come down that column until you get to $n - 1 = 15 - 1$ degrees of freedom. Come across at 14 and there you have it: 2.145, the value of t. Use this in $\mu \pm t\hat{\sigma}_{\bar{x}}$. Though the t table and z table are different, use the t value just as you would the z value in the estimation of means.

Key Cases

What is the optimum number of hooks a key case should have? (Do you suppose the manufacturer actually takes a sample to determine this? What are some of the difficulties involved? How else could he make the decision?)

Let's say he takes a random sample of 20 people on Michigan Avenue in downtown Detroit, asking each person how many keys he is carrying. The sample shows an average of 6.9 with a standard deviation of 2.3 keys. How about a 90% confidence interval?

$$\hat{\sigma}_{\bar{x}} = \frac{s}{\sqrt{n}} = \frac{2.3}{\sqrt{20}} = \frac{2.3}{4.47} = 0.51$$

$$t_{0.050}\hat{\sigma}_{\bar{x}} = (1.729)(0.51) = 1.7(0.51) = 0.87$$

$$\bar{x} \pm t\hat{\sigma}_{\bar{x}} = 6.9 \pm 0.87 = 6.03 \text{ to } 7.77 = 6.0 \text{ to } 7.8 \text{ keys}$$

Good! You estimate the population mean to be between 6.0 and 7.8. When you go into production, you'll have to settle for a whole number of keys —probably 7, or for some cases, 6 or 8. But the confidence interval helps, and t made it possible with such a small sample as 20 cases.

Brown-Tail Moth Pupation Period

The brown-tail moth (Nygmia phaeorrhoae) pupates (goes from the larval stage to an adult while in its cocoon). Timing this pupation period on 15 moths, a student comes up with a mean time of 330 hours and a sample standard deviation (s) of 45 hours. Find an 80% confidence interval estimate of the pupation time.

$$\hat{\sigma}_{\bar{x}} = \frac{s}{\sqrt{n}} = \frac{45}{\sqrt{15}} = \frac{45}{3.87} = 11.63 = 11.6 \text{ hours}$$

$$t_{0.10}\hat{\sigma}_{\bar{x}} = 1.345(11.6) = 15.6$$

$$\bar{x} \pm t\hat{\sigma}_{\bar{x}} = 330 \pm 15.6 = 314.4 \text{ to } 345.6 \text{ hours}$$

The 80% confidence interval estimate for the pupation period for the brown-tail moth is between 314 and 346 hours.

11–7 SUMMARY

The problem of estimating μ is similar to the problem of estimating π; use $\hat{\sigma}_{\bar{x}} = (s/\sqrt{n})$ instead of

$$\hat{\sigma}_p = \sqrt{\frac{p(1-p)}{n}}$$

Go again to Table VIII on the last pages of our appendix. Use it as you work on the exercises.

If the sample size is less than 30, use the t distribution.

Once again, point estimates are almost never exactly right—and we really have no way of telling how close they are. Using confidence intervals furnishes more security, and also a means of expressing our confidence numerically. As usual, some approximations are involved, such as the shape of the distribution and the use of s for σ. But coming up with a confidence interval estimate is certainly a refinement over guessing.

Which do you prefer?

(a) A liquor distributor gets orders for an average of 120 cases of whiskey a day.

(b) A liquor distributor has daily orders for between 98 and 142 cases of whiskey a day (98% confidence interval).

Which one enables you to decide what size delivery truck to use?

We are now about to begin Part III of the four parts of this book. It is called "Statistical Concepts for Testing Hypotheses" and consists of four chapters, the first two using normal distribution directly, the third using chi-square, and the fourth, the analysis of variance. Our first chapter in Part III will ask questions about proportions or counts. For instance, if the liquor distributor claims that he orders 120 cases of whiskey a day, does an average of 111 ordered cases for a sample of 30 days disprove his claim?

• III •

STATISTICAL CONCEPTS FOR TESTING HYPOTHESES

· 12 ·

Testing Hypotheses—Proportions

The shrewd guess, the fertile hypothesis, the courageous leap to a tentative conclusion—these are the most valuable coin of the thinker at work . . . . Yet in many classes in school, guessing is heavily penalized and is associated somehow with laziness.

JEROME S. BRUNER
The Process of Education (1963)

OBJECTIVES

To test an hypothesis (or a claim) about a population proportion by comparing it with the proportion in a random sample from that population.

SYMBOLS AND FORMULAS FIRST USED IN THIS CHAPTER:

H_0 (H sub 0)	the null hypothesis
H_1 (H sub 1)	the alternative hypothesis
α (alpha, the Greek *a*)	the level of significance, the probability of rejecting a true hypothesis, the probability of a Type I error
β (beta, the Greek *b*)	the probability of accepting a false hypothesis, the probability of a Type II error

12–1 EXERCISES

1. Take a sample of newly born guinea pigs to check the genetic theory that 3 out of 15 should be black. Formulate a null hypothesis and an alternative hypothesis.

2. A popularity poll claims that Jones will get 0.55 of the votes. Show a null and an alternative hypothesis.

3. A tire manufacturer claims an average mileage per tire of 30,000 miles. Is the question really just whether he is right or is it whether he has "overclaimed"? Formulate an appropriate null and alternative hypothesis.

4. A car salesman says that not more than one in a hundred of his deluxe models comes in for anything but trivial repairs during the first twelve months after original purchase. From a null hypothesis go on to an alternative hypothesis.

5. If, according to genetic theory, 3 out of 15 guinea pigs will be born black, what would 13 black ones out of a sample of 100 suggest about the theory at a 1% level of significance?

   Always show clearly the null hypothesis you are testing and the alternative hypothesis. Also a graph, showing a normal distribution, the location of decision lines, and the sample statistics, is practically an indispensable conceptual aid. Construct it as you go along, not after you have finished.

6. If the popularity poll claims Jones will get about 0.55 of the votes and a sample of 980 shows 559 favoring Jones, what's your conclusion at a 5% level of significance? Hypotheses clearly again.

7. The dean of students in a large university is informed that 7% of the students say they have taken LSD. A random sample of 200 students showed 5% admitting to having tried LSD. If these students were truthful, what would their response, at $\alpha = 0.01$, indicate about the information given the dean? Make your conclusion stand out from your calculations.

8. The U.S. Forestry Service states that even in the heavily populated North Atlantic Region, as much as 68% of the area is still forest. If a random sample of 200 acres shows 60% wooded enough to be called forest land (something pretty difficult to define precisely), at a 0.01 level of significance, does this discredit the Forestry Service statement?

9. An economist claims that 90% of the value of Venezuela's current exports is oil. How would you get a random sample to check this claim?

10. A random sample of 5000 Californians showed 41% having Type A blood. Make a 99% confidence interval estimate of the percent of Californians having Type A blood.

11. A processor of ground beef states that not more than 2 in 100 of their packages will contain more than 25% fat. A sample of 64 packages was taken at random, and 2 packages were found to contain more than 25% fat. What does this do to the processor's statement at a 5% level of significance?

12. If the IQs of high-school juniors across the country are normally distributed with the theoretical mean of 100 and a standard deviation of 13, what is the likelihood that a student chosen at random will be eligible for a "special" class ($IQ \leq 80$)?

13. A pharmaceutical firm declares that at least 43% of the people questioned preferred the firm's "impregnated" aspirin tablets to any other brand. (Note that there is no statement about sample size, or where the sample was taken, or when it was taken. Note also that the advertiser comes up with an odd percent, to convince you that it actually came from a real sample.) Suppose you took a random sample of 100 aspirin users (a somewhat difficult sample to randomize), asked what aspirin the users preferred, and found that 38% claimed they preferred the "impregnated." What would this suggest about the manufacturer's claim at a 0.02 level of significance?

14. A doctor wants to determine whether a new salve is really better than the one he usually recommends, which he estimates to be 75% effective in curing eczema. He tries it on 30 patients and considers the new salve to be effective in 24 of the cases. What statistically sound conclusion can he draw at a 0.01 level of significance?

15. A medium claims that she has extrasensory perception (ESP). To prove it, she says that, blindfolded, she can tell more often than not whether somebody in the next room draws a black card from an ordinary pack of cards. She is challenged and a stranger draws 1000 times in the next room, replacing the drawn card and reshuffling the pack after every draw. The medium gets 560 guesses right. At a 0.01 level does this suggest that she has ESP?

16. A TV commentator in Connecticut made the statement that 25% of all the cars on the highway do not meet the requirements for a sticker because the horn, headlights, directional lights, brakes, or some other fea-

ture is faulty. An official in the Department of Motor Vehicles thinks this is an overstatement and immediately sets about getting a sample of 400 cars selected at random from different highways at different times of day, finding that 68 cars did not meet the requirements. What could he conclude? Select a level of significance for him.

17. A social scientist claimed that $\frac{1}{2}$ of 1% (0.005) of American adults are alcoholics. (A rather precise proportion considering how imprecisely an alcoholic can be defined.) Suppose it were possible to get a random sample of 900 American people and from this sample 7 were found to be so-called alcoholics. What would this suggest about the scientist's claim at a 0.02 level of significance?

18. A random sample from a day's output of structural steel showed a compressive strength of 63,472 pounds per square inch with a standard deviation of 9582 pounds per square inch ($n = 30$). Construct a 99% confidence interval estimate for the running daily average of the compressive strength of this structural steel.

12–2 FORMING HYPOTHESES

Our estimation of population parameters was based on sample statistics. From these sample statistics we came up with confidence interval estimates of population proportions or population means. However, often we will have some hypothesis about a population parameter to start with, and will want to take a sample to test the reliability of that hypothesis. This is the type of statistical decision problem we shall now attack: testing hypotheses about population parameters by the use of sample statistics.

The word *hypothesis* is, pretty obviously, a combination of *hypo* and *thesis*: *hypo*, below; *thesis*, to put. An hypothesis is an assumption we put down for our reasoning to stand on. If you hypothesize that honesty is always the best policy, then you reason from it that it behooves you to tell the truth. If you believe the claim that your chances of survival in a bad car crash are much better if you wear both seatbelts, then it would be reasonable to use them—if you believe in survival, too. If you believe, also, that the circumstances promoting a bad crash are unpredictable, then you will wear both belts whether you drive around town or on the highway.

You can tell the hypotheses upon which a reasoning person operates if you study his actions. The results of a me-first philosophy are not very difficult to detect. A you-first philosophy is not harder to detect, but harder to find. Even those people who perform so erratically that there appears to be no consistent philosophical hypothesis whatever behind their actions may be said to stand upon the hypothesis that reasoning really doesn't matter, anyway—impulse is good enough.

Geometry since 300 B.C. has been the classic example of deductive reasoning, proceeding from hypothesis to conclusion by the rules of logic. Using axioms and theorems, logical progress is made from, for instance, the hypothesis that the sum of the interior angles of a triangle is equal to one straight angle to the conclusion that the sum of the exterior angles (on the same side) is equal to two straight angles. This rigid kind of mathematical logic fascinated philosophers, luring them into creating axioms and theorems about life from which logically to draw conclusions—Spinoza, Kant, Descartes, Thomas Jefferson—about religion, morals, and politics. (See the American Declaration of Independence.)

But when you get out of the world of abstraction, you find that you can't really prove hypotheses to be true. That the sum of the interior angles of a triangle is equal to one straight angle is fairly easy to prove. The hypothesis in economics that demand always increases supply can't be proved. You can only say that it is usually true. And you'll need statistical concepts and techniques in analyzing samples to tell how usual (how probable) it is that the hypothesis is true. That is the problem before us: to test hypotheses, accepting or rejecting them, and stating statistically the probability that our decision is right or wrong.

We will formulate hypotheses from factual situations: 9 out of 10 children like peanut butter; transportation in the United States takes up 23% of all energy demands; Wearwell tires will last at least 30,000 miles; 4 out of 5 American women prefer Snowhite toothpaste over any other brand; half the accidental deaths in the United States are due to motor vehicle accidents; only 0.25 of Chillfrost refrigerators need adjustment or repairs within the first three months after purchase.

Such statements will be the basis for our null hypotheses. A **null hypothesis** should assume something to be true; it should contain an equal sign ($=$). You can't draw a distribution using π (or μ) as less than or greater than. The distribution has to have a specific central value, which is our null hypothesis. The alternative hypothesis, on the other hand, may be expressed in any one of three different ways: not equal to ($\neq$), less than ($<$), more than ($>$).

Now it is very important to see that when we reject an hypothesis, we are implying another, which we shall call the **alternative hypothesis.** Failure to spell out an alternative hypothesis is a great stumbling block for students. They fail to state an alternative hypothesis, so that when they reject the null hypothesis, no one knows what they do accept. Take the following examples. Say, 9 out of 10 children like peanut butter; a reasonable alternative hypothesis seems to be that this simply is not true, that an average of fewer or more than 9 out of 10 children like peanut butter. Or say, Wearwell tires will last 30,000 miles; the single appropriate alternative hypothesis seems to be that Wearwell tires do *not* last that *long*; therefore reject the null hypothesis only when the mileage is less—if the tires last longer, the consumer won't argue. See if you can construct an alternative hypothesis which the other statements two paragraphs back seem to imply (energy for transportation, etc.). Accurately constructing hypotheses is very important in this course. It is very important in life, too.

As a difficult philosophical example, Napoleon Bonaparte remarked upon reading Laplace's great *Mécanique Céleste*, "You have written a huge book on the system of the world without once mentioning the author of the universe." To which Laplace replied, "Sire, I had no need of that hypothesis." So what could have been Laplace's alternative hypothesis?

Color TVs

We are told that in a certain New Jersey suburban town 80% of the TVs are color sets. We don't know whether this is right or not. Wanting to know, but not wanting to take the time to check every set in town, we decide to attempt to test this hypothesis by a random sample of 100.

12–3 DISTRIBUTION OF SAMPLE PROPORTIONS, AGAIN

Now you see that this is a binomial situation; the TVs are either color sets or not. Since both $n\pi$ and $n(1 - \pi)$ are more than 5, we can risk using a normal distribution as a mathematical model. The hypothetical parameter is $\pi = 0.80$.

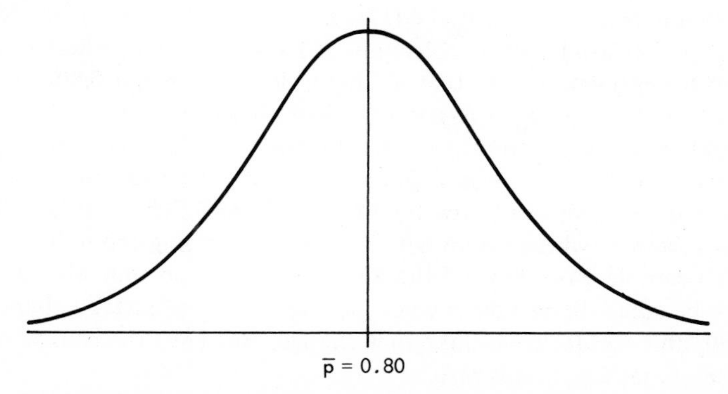

Distribution of sample proportions of color TV sets ($n = 100$, $\pi = 0.80$)

We know from past experience that when a random sample is taken, we can't expect the proportion in it very often, or perhaps ever, to be exactly π. The proportions of color TV sets in more samples will certainly vary, most of them probably coming fairly close to 0.80. but some coming somewhat farther away, and a few a lot farther away.

12–4 DECISION LINES

The question now becomes: how far below or above $\bar{p}$ has the sample proportion got to be for us to reject the hypothesis—0.79, 0.76, 0.88, 0.93? We do have to draw the line somewhere and relying on common sense or intuition is going to have different people drawing lines in different places. And we have already suggested the unreliability of common sense in probability matters. (For instance, would you believe that the probability of drawing any specified poker hand—five exactly specified cards—is about one chance in $2\frac{1}{2}$ million? Is that common sense?)

We shall reject the color TV hypothesis only if our sample p is either too large or too small. But notice that we always assume, before we take the sample, that we have nothing yet to justify our not accepting it. Having *nothing* to make us reject this hypothesis to begin with is why this hypothesis is traditionally called the *null hypothesis*. Symbolically, H_0: $\pi = 0.80$; the null hypothesis (H sub zero) is that π equals 0.80.

If the sample p is significantly larger or smaller than 0.80, then we are going to reject the null hypothesis. But, if you think about it again, rejecting an hypothesis means accepting another. If we reject the null hypothesis that 80% of the TVs are color sets, we are simply accepting the hypothesis that 80% is not true. If so, the alternative hypothesis (H sub 1) may be written H_1: $\pi \neq 0.80$ (π does not equal 0.80). An alternative hypothesis could also have been that π is really more than 0.80 (H: $\pi > 0.80$) or that π is really less than 0.80 (H_1 : $\pi < 0.80$). In our problem about TV sets, it seems a better interpretation of the problem to say that π simply does not equal 0.80—it may be more or less. An alternative hypothesis must always be stated.

Sketch again a theoretical distribution of sample means for our null hypothesis, using $\bar{p} = 0.80$ ($\pi = \bar{p}$) and $n = 100$. We want to decide where to draw our decision lines, how far they are from $\bar{p}$. These lines will show, going out from $\bar{p}$, where we are going to begin rejecting the null hypothesis because a sample p out beyond there from $\bar{p}$ is just too unlikely. Of course, we can't be sure: the sample p could be way out there just by chance, with $\bar{p} = 0.80$, But we decide to take that chance. But keep the chance of being wrong small, say 2% in this case.

If we want our sample proportion to fall outside our decision lines only 2% of the time, we must then draw our decision lines $z\sigma_p$ away from the mean. ($z = 2.33$. Remember?) This 2% is called the **level of significance** of our test. It shows the probability of rejecting a true null hypothesis. It is indicated by a small Greek "*a*" called alpha (α —an eight with no bottom, lying down). ($\alpha = 0.02$.) The decision lines in this case are really $z_{\alpha/2}\sigma_p$ from $\bar{p}$.

$$H_0: \pi = 0.80$$
$$H_1: \pi \neq 0.80$$
$$\alpha = 0.02$$

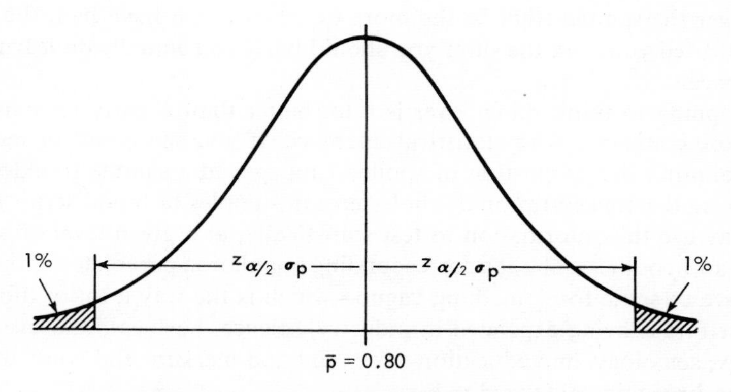

Distribution of sample proportions (n = 100, α = 0.02)

We can say now that we will reject our assumption, $\pi = 0.80$, if the sample proportion falls outside these lines. But this rejection is unlikely—in fact, only 2% probable. Look carefully and you will see that if we use these "decision" lines, then 2% of the time we may be rejecting the null hypothesis when it is true. It's possible, but only 2% probable, that with $\pi = 0.80$ a single sample proportion will fall beyond these decision lines. This is the calculated risk!

You don't want to be wrong? Well, if you push the decision lines farther apart, you will reduce the possibility of rejecting the null hypothesis when it really is true, but—and it is a big "but"—you will increase the probability of accepting H_0 when it is not true.

This dilemma is a common one: drawing a line, which you often have to do to make a decision, always allows the possibility of being wrong. In most of life's situations you pull up your courage and say intuitively, "Here I will draw the line." In statistical inference you have the added assurance of being able to say, "If I draw the line here, I can state the calculated risk."

The businessman, as well as the scientist and engineer, wants something better than intuition or common sense. So we give him something better. In the color TV case, we say that we are willing to take a 2% chance of rejecting the true null hypothesis. With our decision lines, if π really does equal 0.80, we will be rejecting this true null hypothesis only 2% of the time. This is a lot more informative than saying, "Something tells me this is where we should draw the line."

There are situations in life where you and I strongly wish to know where to draw the line (use your imagination). If it is a new experience, we have to rely on intuition, reasoning, or, perhaps, what we have heard from others. If we have encountered the experience before, we can judge somewhat from the samples we have already taken. Theoretically, as in statistical inference, the larger the sample (that is, the more experience you have had, the more sophisticated you are), the surer you should be, if you actually do learn from experience.

Stopping to think things over is a lot better than a gutty response but still loose compared with statistical inference. If you can count or measure (like counting the proportion of spoiled cabbages in a sample truckload or measuring the concentration of cholesterol in samples of blood serum), then you may use this information to test statistically, at a given level of significance, an hypothesis about a corresponding population parameter. Otherwise you have to settle for something vague—which is the way a lot of life runs. Much effort is being expended in industry, science, business, and even psychology, sociology, and education—to count and measure and come up with decision lines and calculated risks.

12–5 LEVEL OF SIGNIFICANCE, TYPE I ERRORS

The likelihood of rejecting a true hypothesis is called a Type I error, the *level of significance*. The level of significance is the likelihood of making a Type I error.

When we decide to draw our decision lines so that if H_0 is true, we will erroneously reject it only 2% of the time (1% above, 1% below), we need to know the distance of the decision lines from $\bar{p}$. And, of course, we need z again in order to get $z\sigma_p$.

From our work on confidence intervals you should recognize that the z we want for 2% is 2.33: This z will leave us 1% on each side. From our work on the distribution of sample proportions (Chapter 8), you should also remember that $\sigma_p = (\sqrt{\pi(1 - \pi)/n})$.

$$\sigma_p = \sqrt{\frac{\pi(1-\pi)}{n}} = \sqrt{\frac{(0.80)(0.20)}{100}} = \sqrt{\frac{0.16}{100}} = \sqrt{0.0016} = 0.04$$

So

$$z\sigma_p = (2.33)(0.04) = 0.093.$$

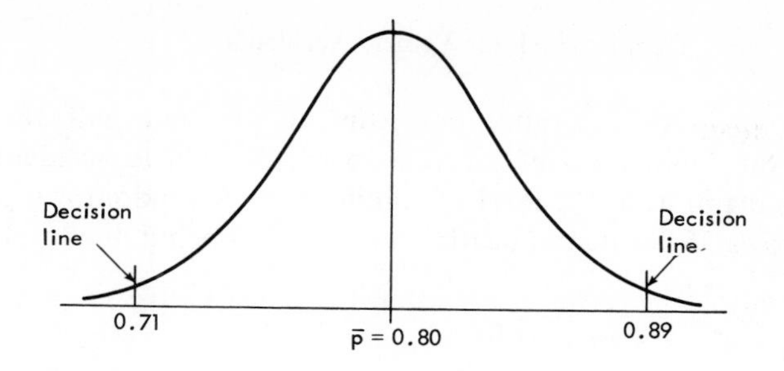

Distribution of sample proportions ($n = 100$, $\sigma_p = 0.04$, $\alpha = 0.02$)

This picture shows us where to draw the lines when $\alpha = 0.02$. We could have chosen $\alpha = 0.50$ ($z = 0.67$), which would mean we were willing to reject a true hypothesis half the time—willing to be wrong half the time. Not a very safe stand to take; recommended calculated risks should be smaller, like those going with the z's we used in estimation. They are listed under "Symbols and Formulas," on the second page of Chapter 10—or have you already memorized them?

Suppose we draw the random sample of 100 TVs and find 68% to be color sets. The sample proportion falls below the lower decision line. This

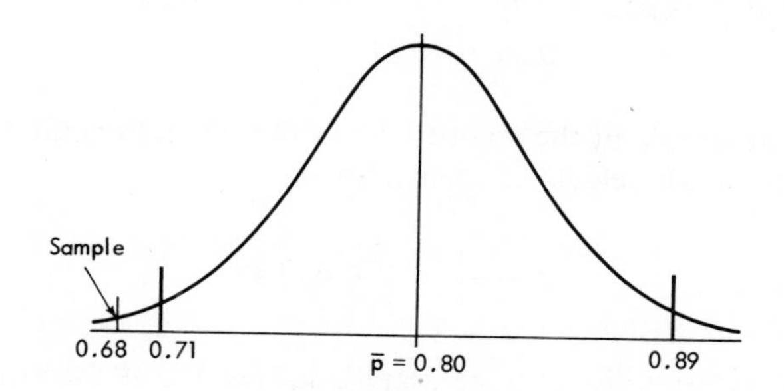

Distribution of sample proportions ($n = 100$, $\sigma_p = 0.04$, $\alpha = 0.02$)

result is *so* improbable by chance when $\pi = 0.80$ that we reject $H_0: \pi = 0.80$ on this basis. The sample suggests only that $\pi = 0.80$ is probably wrong because p falls in a rejection area. So we accept the alternative hypothesis ($H_1: \pi \neq 0.80$) and hope the sample has been representative.

Reject H_0: the proportion of color TV sets is probably not 0.80.

Deaths by Motor Vehicle Accidents

An insurance company claims that half the accidental deaths in the United States are due to motor vehicle accidents. Test the company's claim at a 2% level of significance. A random sample from the records shows 56 accidental deaths out of 115 as being due to motor vehicle accidents.

$$H_0: \pi = 0.50$$
$$H_1: \pi \neq 0.50$$
$$n = 115$$
$$\alpha = 0.02$$
$$z_{\alpha/2} = 2.33$$
$$\sigma_p = \sqrt{\frac{\pi(1-\pi)}{n}} = \sqrt{\frac{(0.50)(0.50)}{115}}$$
$$= \sqrt{\frac{0.250}{115}} = \sqrt{0.00217}$$
$$= 0.0466$$
$$z_{\alpha/2}\sigma_p = (2.33)(0.0466) = 0.1086 = 0.11$$
$$p \pm z_{\alpha/2}\sigma_p = 0.50 - 0.11 \text{ to } 0.50 + 0.11$$
$$= 0.39 \text{ to } 0.61$$

A random sample of the records shows that 56 accidental deaths out of 115 are due to motor vehicle accidents. So

$$p = \frac{x}{n} = \frac{56}{115} = 0.49$$

Place the decision lines on the graph and then locate the sample proportion on the graph. See graph at top of facing page. The sample proportion is well within the acceptance area.

Do not reject H_0: the proportion of accidental deaths due to motor vehicle accidents is probably 0.80.

This doesn't prove H_0 to be true. There simply are not grounds for rejecting it. Not rejecting the null hypothesis never actually proves it to be true.

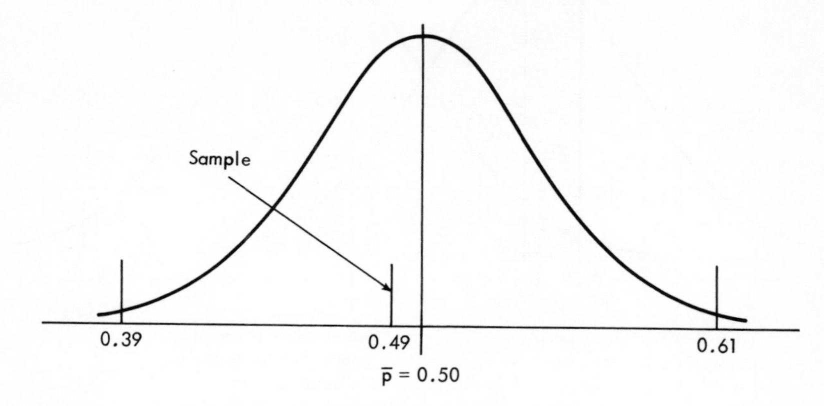

Distribution of sample proportions for vehicular deaths ($\sigma_p = 0.0466$, $n = 115$, $\alpha = 0.02$)

In a scientific approach to a study of nature, there are no proofs as there are in abstract mathematics. The information from samples simply either doesn't contradict a given hypothesis or it does contradict it. One contradiction can destroy a scientific theory; no amount of supporting information can ever prove it.

This stance on the impossibility of concrete proof in science was perhaps first clarified by the great English philosopher David Hume (1711–1776). There is a parallel in statistical inference where a null hypothesis is either not rejected or an alternative hypothesis is accepted tentatively. Nothing is proved.

12–6 TYPE II ERRORS

Making a Type II error is accepting a null hypothesis when that hypothesis is really false. Our previous graph showed the distribution of sample proportions with decision lines for $\alpha = 0.02$. We were to accept the null hypothesis (H_0: $\pi = 0.50$) when the sample proportion fell between 0.39 and 0.61.

But what is the probability, then, of accepting this null hypothesis when it really isn't true? Stop! You can't answer that. You need to know the true population proportion in order to get the numerical probability of a Type II error, and it would be rather superfluous to draw decision lines to check the null hypothesis about π when you know already what π is.

On the other hand, you can answer this type of rather useful question: Suppose we use this decision plan and suppose that the true π is really 0.45 instead of 0.50. What would be the probability of accepting the false null hypothesis that $\pi = 0.50$?

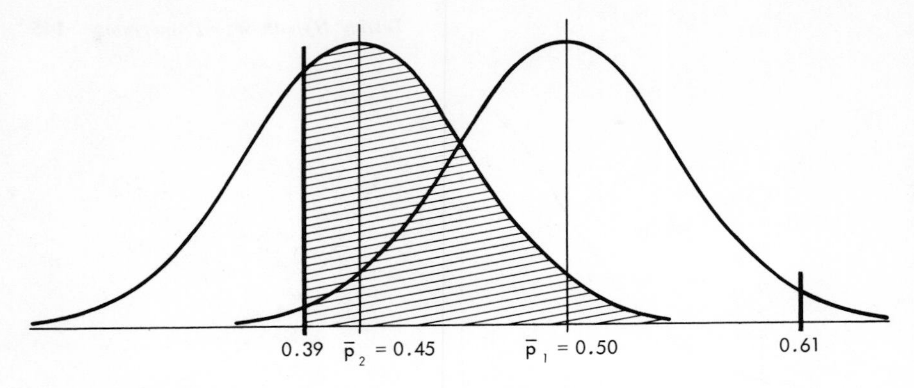

The curve for $n = 115$, $\pi = 0.45$ is laid on the same horizontal axis as for the $\pi = 0.50$ distribution; our decision lines remain as proposed. Bearing in mind that the true distribution is now $\pi = 0.45$, we see at once that the probability of accepting H_0: $\pi = 0.50$ is that of the shaded area above our lower decision line. This is the probability of accepting $\pi = 0.50$ when actually $\pi = 0.45$. This is the probability of accepting a false hypothesis, a Type II error. It is symbolized by the Greek β (beta).

The σ_p for $\pi = 0.45$ will actually be different from what it was for $\pi = 0.50$. To get the numerical probability of a Type II error, you will have to do some calculating.

$$\sigma_p = \sqrt{\frac{\pi(1 - \pi)}{n}} = \sqrt{\frac{(0.45)(0.55)}{115}}$$

$$= \sqrt{\frac{0.2475}{115}} = \sqrt{0.00215}$$

$$= 0.0464$$

$$z = \frac{0.39 - 0.45}{0.0464} = \frac{0.06}{0.0464} = 1.29$$

$$\text{Area} = 0.4015$$

This is the acceptance area between $\bar{p}_2$ and the lower decision line 0.39. To the right of $\bar{p}_2$ lies 0.5000 more of the area (that rejection area beyond 0.61 now being negligible).

$$P_{(\text{Type II error})} = 0.4015 + 0.5000$$
$$= 0.9015$$
$$\beta = 0.902$$

Pretty high in this case! If the π is really 0.45, we are very likely, using the original plan, to accept the false hypothesis.

12–7 ONE-SIDED TESTS

Snowhite Toothpaste

A manufacturer claims that 4 out of 5 American women prefer his Snowhite toothpaste over any other brand. The manufacturer knows that numbers always give the appearance of reliability. But don't *you* be taken in. No mention is made of the sample size. Was it just 4 out of 5 women, or was it 40 out of 50, or 80 out of 100? Nor is there any mention of *how* the sample was taken. Were the women all relatives of the manufacturer? Were they all from the same neighborhood? Or were they chosen randomly from different streets in different cities in different states? And what about the ages of the women and their economic status? As you already know, if the sample is not *really* representative of the population, your statistics will be misleading no matter how carefully you calculate them.

The hypothesis is that 4 out of 5 women prefer Snowhite. Four out of 5 again means 80%, or $\pi = 0.80$; this is an hypothetical parameter. We don't know what the manufacturer actually based his claim on, but we will check it with a random sample of our own. The larger the sample, of course, the greater the expense in time and money. We take a reasonably small (?) sample and, hopefully, a representative one. We use $n = 100$ (although $n = 1000$ would be more reliable). Since both $n\pi$ and $n(1 - \pi)$ are more than 5, we can risk counting on a normal distribution of the sample proportions.

The next sketch is for a theoretical distribution of sample proportions where $n = 100$ and $\pi = 0.80$. Since the mean of the sample proportions would equal the population proportion, use $\bar{p}$ for π.

But when you think about it, isn't it more reasonable to check only whether the producer's claim is too high? If it's too low, you don't care because the toothpaste is even better than it's claimed to be. When he says 4 out of 5 women prefer Snowhite, doesn't he mean that as large a proportion as 0.80 do, or even more? You really want to check whether he has overstated his case.

We make $H_1 : \pi < 0.80$. We are going to reject H_0 only when p is significantly ($\alpha = 0.01$) lower than π. What we want to find out is whether the sample shows that the manufacturer has exaggerated women's preference for Snowhite. So we draw a line on the left side of the graph only, leaving all 1% of the area to the left of the decision line. We reject only when the sample is too low, that is, to the left of the decision line. This, again, means that the probability of a Type I error is 1%.

In the sample, p comes out to be 0.68. Is this proportion enough smaller than 0.80 to warrant rejection? The decision line must lie at $\bar{p} - z\sigma_p$; all the 1% rejection area is on that one side of the graph.

$$\sigma_p = \sqrt{\frac{\pi(1-\pi)}{n}} = \sqrt{\frac{(0.80)(0.20)}{100}}$$

$$= \sqrt{\frac{0.16}{100}} = \sqrt{0.0016}$$

$$= 0.04$$

$$\bar{p} - z\sigma_p = 0.80 - 2.33(0.04)$$

$$= 0.80 - 0.09$$

$$= 0.71$$

The following graph answers our question.

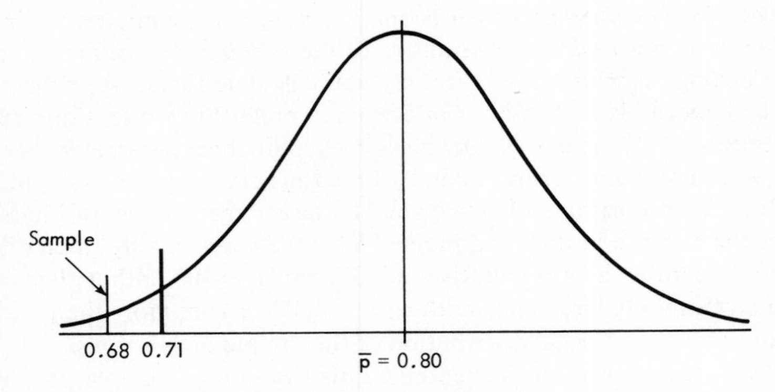

Distribution of sample proportions (n = 100, π = 0.80)

The sample proportion falls below our decision line, so again we reject H_0.

> Reject H_0: the proportion of women tooth brushers who prefer Snowhite is probably less than 0.80.

Refrigerator Repairs

A manufacturer claims that only 0.25 of his refrigerators need adjustments or repairs within the first three months. (Do you find this claim inviting?) A consumers' association found that 55 out of a random sample of 230 needed some kind of adjustment or repair within the first 3-month period. At a 0.05 level of significance, what does the sample proportion suggest?

You will consider the manufacturer's claim untrue only if the sample shows a significantly *higher* proportion of refrigerators that need adjustments or repairs. You do not intend to dispute his claim if the sample shows fewer refrigerators needing repair than he claims.

$$H_0: \pi = 0.25$$
$$H_1: \pi > 0.25$$

Making the two hypotheses reveal what you want to ask about the situation is an absolutely essential first step in the decision process. You must know what the information really says, and you must know what you are going to ask about it. If the problem can be reasonably interpreted in different ways, then different reasonable solutions must be allowed. The classroom task of getting "the right answer" is not strictly applicable here.

I repeat, there is no moving toward a solution until you pose a problem. In this case, you have decided to ask whether the sample suggests that the proportion of refrigerators needing repairs in the first three months is really more than 0.25.

Compute p first; if it happens to come on the side of $\bar{p}$ opposite from your decision line (in this case, to the left of $\bar{p}$) there is no need for going any further. You can't possibly reject π; your rejection area is over on the other side. So

$$p = \frac{x}{n} = \frac{55}{230} = 0.239 = 0.24$$

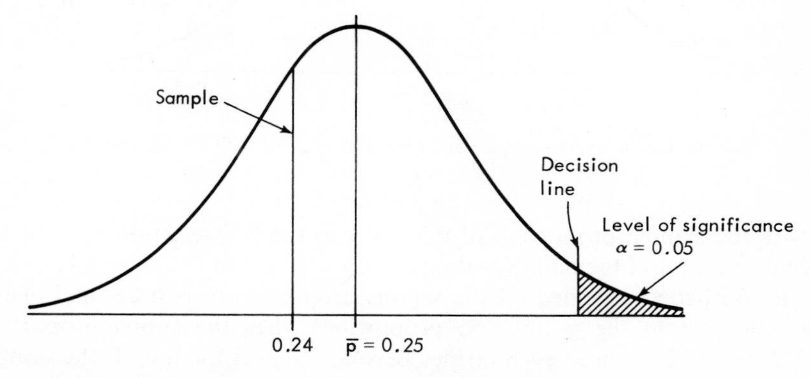

Distribution of proportions for refrigerators needing repairs (n = 230)

Yes, indeed. You can stop right there! p is on the good side of $\bar{p}$.

Now, suppose that the number of refrigerators that needed repair had been 72.

$$p = \frac{72}{230} = 0.313$$

This is to the right of $\bar{p} = 0.25$.

Is it far enough to the right to be significant? Here you go.

$$\alpha = 0.05$$
$$n = 230$$
$$H_0: \pi = 0.25$$
$$H_1: \pi > 0.25$$
$$\sigma_p = \sqrt{\frac{\pi(1-\pi)}{n}} = \sqrt{\frac{(0.25)(0.75)}{230}}$$
$$= \sqrt{\frac{0.1875}{230}} = \sqrt{0.000815} = 0.0286$$
$$z\alpha\sigma_p = (1.64)(0.0286) = 0.0469 = 0.05$$
$$\bar{p} + z\sigma_p = 0.25 + 0.05 = 0.30$$

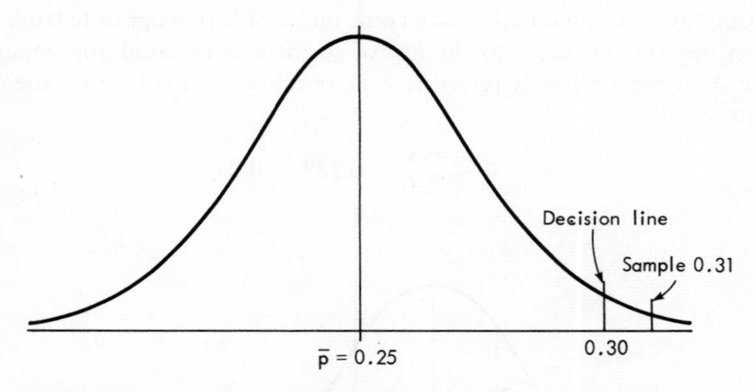

Distribution of sample proportions ($n = 230$, $\sigma_p = 0.0286$, $\alpha = 0.05$)

As the sample proportion of 0.31 falls in the 5% rejection area to the right of 0.30, reject the null hypothesis.

If you hadn't rounded off the sample proportion to two decimal places (the precision in the population proportion), then the sample proportion (0.313) would have been even farther beyond the decision line. If the sample proportion had come very close to the decision line, you should not go back and calculate more closely; instead, take another sample.

> Reject H_0: the proportion of refrigerators needing adjustment in the first three months is probably more than 0.25.

12–8 SUMMARY

Repeat: When p and $z\sigma_p$ are very close, don't go back and carry out calculations to more decimal places or greater accuracy. Take another sample.

This numerical hypothesis testing gives you a somewhat justifiable sense of security. It is a powerful concept, and you can imagine its very practical importance in a great many areas of work. If $n\pi \geq 5$ and $n(1-\pi) \geq 5$ (that is, if the sample is not too small, nor the proportion too far from 0.500), the distribution of the proportions will be approximately normal, and all you will need to be given to test the hypothesis about π are n and α! To get so much help out of so little data is a handsome return on your investment.

You will even be more secure if you study Testing Hypotheses, Proportions, in Table VIII of the appendix (page 243). This tells you how to draw decision lines, given a level of significance (that is, the probability of risking a Type I error).

You were given an introduction to Type II errors, but their probabilities will not be questioned in the exercises.

Never attempt to test a hypothesis without a specified level of significance. That is the statistical way. In everyday life you can say, "Those differences don't mean anything," and you can get away with it. You can't get away with it in statistical inference. You've got to have a level of significance. When someone asks, "Are these differences significant?" you don't ever answer right off; you always ask first, "At what level of significance?" If you don't get an answer, you don't give an answer. This is an important concept to be learned from this course.

In statistically testing hypotheses about proportions, form a null hypothesis and an alternative hypothesis. If the sample proportion is outside the limits where the level of significance tells you to draw your decision line or lines, reject the null hypothesis in favor of the alternative hypothesis. Otherwise do not reject the null hypothesis.

In a court trial the defendant is always assumed to be innocent until proved guilty. The null hypothesis is that there is yet no evidence to show that the defendant is guilty. The alternative hypothesis, obviously, is that the defendant *is* guilty. The defendant is convicted only when the court has grounds to reject the null hypothesis that he is innocent. The level of significance is somewhat vaguely defined as "beyond a reasonable doubt."

Court adjourned!

• 13 •

Testing Hypotheses—Means

I frame no hypotheses; for whatever is not deduced from the phenomena is to be called an hypothesis; and hypotheses, whether metaphysical or physical, whether of occult qualities or mechanical, have no place in experimental philosophy.

SIR ISAAC NEWTON
Letter to Robert Hooke (1675)

OBJECTIVES

To test hypotheses about population means by comparing them with the means of random samples from those populations.

13–1 EXERCISES

1. Explain carefully how Newton's use of "hypotheses" differs from ours.

2. A computerized airline ticket reservation system checks out reservations in a mean time of 46 seconds. A newly devised computerized system checks out 100 reservations in an average time of 40 seconds. Frame a null and an alternative hypothesis to help the airline management decide whether to change systems.

3. A research agency claims that the average income in Millville is $8000. A random sample of 200 incomes shows an average of $8100. Formulate a null and an alternative hypothesis. Suggest conditions under which a one-sided test might be desirable.

4. A fireworks manufacturer claims the mean burning time for its skyrocket fuses is 10 seconds. A random sample of 50 is drawn from a large shipment of these fuses in an effort to determine, at a 0.05 level of significance, whether the shipment meets the manufacturer's claim. What will be your conclusion concerning the claim if the mean burning time of the sample is 9.3 seconds with a standard deviation of 0.5 seconds?

5. It is assumed that the melting point of lead is 327°C. If a metallurgist makes the same test 36 times and gets a mean melting point of 325°C with a standard deviation of 2°C, what would you conclude, with a 0.01 level of significance, about the hypothesis that the melting point is 327°C?

6. A manufacturer wants to buy a new stitching machine to replace the old one, which put out 200 socks an hour on the average. He tested a new machine for 30 hours, found it averaging 215 socks per hour with a standard deviation of 11 socks. At a 0.02 level of significance, is the new machine better? What is the probability of a Type I error?

7. A furniture retailer claims that its mean overdue charge account is $76.25. A retail credit association wants to check this statement. It takes a sample of 144 overdue accounts, selected at random from the company's files, and finds the sample average to be $81.23 with a sample standard deviation of $9.30. What conclusion, at a 0.05 level of significance, can the association draw about the retailer's statement?

8. Investigating an alleged unfair trade practice, a statistician working for the Federal Trade Commission takes a random sample of 50 "8-ounce" packages of marshmallows, getting a mean of 8.13 ounces and a standard

deviation of 0.10 ounces. At a 0.05 level of significance, test whether the sample evidence indicates unfair practice.

9. A manufacturer claims that the breaking strength in pounds per square inch of a particular type of ceramic vacuum tubes is 110 or better. If, in a sample of 100, a buyer finds a mean breaking strength of 105 pounds per square inch with a standard deviation of 15.9 pounds per square inch, what, at a 0.01 level of significance, should he conclude about the vendor's claim? What is the probability of a Type I error? When would a Type II error occur? What is the probability of a Type II error?

10. A cereal manufacturer decides to "fortify" his barley flakes by adding raisins. In a $1\frac{1}{4}$-ounce box of barley flakes, he wants to have a dozen raisins. The raisins are added by a machine on the packaging line. What would be a 99+% safe guess as to the variation in raisin count in samples of 50?

11. The theoretical IQ mean is 100. Suppose we check this hypothesis by taking a random sample of 1000 people. (Where and how would you get this random sample?) The sample shows $\bar{x} = 99$ and $s = 12$. Would you reject the hypothesis that $\mu = 100$ at a 0.02 significance level? (We use μ_x or just μ for the population mean.)

12. The *control chart* is a basic statistical concept used in quality control in industry, business, and many other areas.

    Here is a control chart example. A machine cuts $\frac{1}{4}$-inch steel rods into pieces 2 inches long. When cutoff machine is running properly, the standard deviation of these lengths is 0.005 inches. In what interval would you expect to find practically all the means of samples, $n = 30$? (For "practically all" use $z = 3$.) When a sample mean exceeds these interval boundaries, the machine is to be shut down and adjusted. A basic objective in quality control is to avoid having to sort out defective pieces by keeping the manufacturing process within the 3-sigma limit of variation. When a sample of the production process is taken in this case, what is the null and what is the alternative hypothesis?

13. Radioactive iodine is claimed to be more than 75% effective in relieving patients from the almost intolerable pains of angina pectoris. Tested in 50 cases, it was declared effective in 35 cases. How would this sample statistic affect the claim if $\alpha = 0.02$?

14. The U.S. Bureau of Land Management used a random sample of 120 tree diameters ($\bar{x} = 18.6$ cm and $s = 3.6$ cm) to estimate a 0.95 confidence interval for the mean tree diameter of a wooded area. Try it.

15. A new achievement test was given to 625 randomly selected students from the many thousand freshmen entering junior colleges throughout the state of California. If they averaged 498 with a standard deviation of 16, how sure could you be that a single California student randomly selected from freshmen entering junior colleges would have a score between 450 and 546?

16. An ichthyologist (a specialist in the study of fish) states that his research shows the average length of adult North Atlantic flounders to be 91 centimeters with a standard deviation of 9 centimeters. Make a 95% confidence interval estimate of the mean length of these flounders in a random sample of 50.

17. The Joneses are going to deep Africa with their infant son and want to be sure that they have enough diapers for a year. If each diaper has a mean "life use" of 96 days and a standard deviation of 22 days, assuming a normal distribution, how many diapers should the Joneses pack for a 0.90 chance of having enough to use 10 a day for a year?

18. The Monadnock Insurance Company spent $225 million building the highest building in a certain city. Officials stated that one of their objectives was to have at least 75% of the people mention the Monadnock Building when asked for the most conspicuous feature of the city skyline. Asking about the most prominent feature of the city skyline, an investigator took a random sample of 1000 answers in hotels, subways, stores, and on the streets. Of this sample, 720 named the Monadnock Building. Should this, at a 5% level of significance, satisfy the officials?

13–2 TESTING HYPOTHESES ABOUT POPULATION MEANS

The concept of hypothesis testing we have thoroughly discussed. This chapter should be fairly clear sailing for you. Going from hypothesis testing about proportions to hypothesis testing about means involves a similar statistical concept with a different solution technique. Whereas before we used

$$\bar{p} = \pi, \qquad \sigma_p = \sqrt{\frac{\pi(1-\pi)}{n}}, \qquad \text{and } z \text{ for } \alpha \text{ or } \frac{\alpha}{2},$$

we will now use

$$\mu_{\bar{x}} = \mu, \qquad \sigma_{\bar{x}} = \frac{\sigma}{\sqrt{n}}, \qquad \left(\text{or } \frac{s}{\sqrt{n}} \text{ if } \sigma \text{ is not given}\right),$$

$$\text{and } z \text{ or } t \text{ for } \alpha \text{ or } \frac{\alpha}{2}$$

It is fairly common practice to use μ for the population mean and σ for the population standard deviation, though we could have used the more explicit symbols μ_x and σ_x for the same parameters.

We again have the situation of the distribution of a sample statistic, the mean in this case. If the sample size is equal to or more than 30, we can use our z table for the standard normal distribution—a candle in the darkness of uncertainty to light our way (see Arbuthnot, title page of Chapter 10). Form a null and an alternative hypothesis and use decision lines to test the null hypothesis at a specified level of significance; if the null hypothesis is rejected, accept the alternative hypothesis.

A few examples will familiarize you with the technique for constructing your graph of the distribution of sample means with its decision lines. Meanwhile the concept of hypothesis testing should be coming even more sharply into focus. What are we doing—finding the truth? No. We are just checking with our sample information. We never really prove our hypotheses; we simply find out whether we have grounds for rejecting one and accepting the other.

Rejecting an assumed hypothesis simply suggests accepting an alternative hypothesis that the true parameter is probably not the assumed one, or that the true parameter is probably more than the assumed one, or that the true parameter is probably less than the assumed one. Just one of these three important statements must be stated as an alternative hypothesis.

The graph we draw is a probability distribution, showing the mean value as most likely, the extreme values as least likely. The precision of our results is predetermined by the precision of the given information. These units of precision would seem to freeze the distribution into discrete steps. Having chosen, say, two decimal places of precision, this would amount to having tiny steps $\frac{1}{100}$ of a unit wide. Appreciating some of the other approximations involved, however, we take the liberty of simplifying matters by drawing a smooth, continuous curve. Think it over.

Spray-Paint Drying Time

A paint manufacturer claims that his industrial paint sprayed on iron castings will dry in 3 minutes with a standard deviation of 1 minute. (Manufacturers usually don't state standard deviations for any specifications but it would certainly be more revealing if they did. If we don't have σ for drying time, we'll have to use the s from our sample.) A plant manager tries the paint on 300 castings and finds an average drying time of 3.2 minutes. Should he reject the manufacturer's claim at $\alpha = 0.01$?

We shall use a one-sided test because we are thinking that the plant manager is not at all concerned about the paint drying faster than is claimed. This is our assumption; it may not be the most appropriate one. But unless we have someone to ask who is personally involved in the problem situation, we can only read the verbal presentation very carefully ourselves and select what we think are the most appropriate hypotheses.

$$H_0: \mu = 3 \text{ minutes (because } \mu_{\bar{x}} = \mu)$$
$$H_1: \mu > 3 \text{ minutes (one-sided)}$$
$$n = 300$$
$$\sigma = 1.0 \text{ minutes}$$
$$\sigma_{\bar{x}} = \frac{\sigma}{\sqrt{n}} = \frac{1}{\sqrt{300}} = \frac{1}{17.32} = 0.0577$$
$$\alpha = 0.01, \qquad z = 2.33$$
$$\mu + z\sigma_{\bar{x}} = 3.0 + 2.33(0.0577)$$
$$= 3.0 + 0.134 = 3.134 = 3.13$$

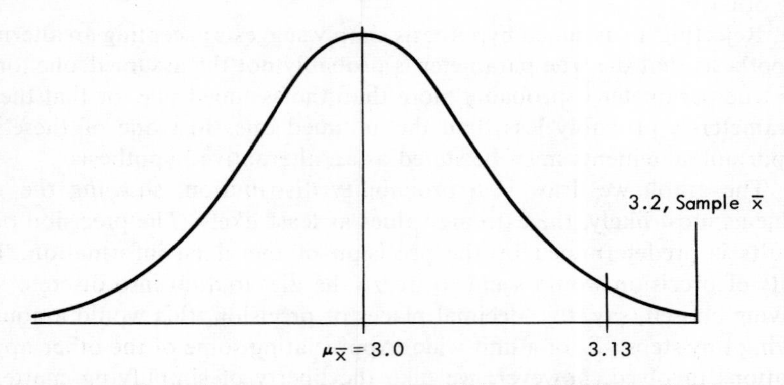

Distribution of sample means in minutes ($n = 300$, $\sigma_{\bar{x}} = 0.0577$, $\alpha = 0.01$)

The graph shows quite clearly that we are going to reject the manfacturer's claim. With a population standard deviation of 1 minute, it looked at first as though $\bar{x} = 3.2$ minutes might not be too bad. But when we went over to the standard deviation for *sample means*, we had to divide the population standard deviation by the square root of 300; this produced a much smaller $\sigma_{\bar{x}}$ than the σ ($\sigma = 1.0$ minute to $\sigma_{\bar{x}} = 0.0577$ minute) and consequently pulled the decision line in closer, increasing the rejection possibility.

> Reject H_0: the paint drying time appears to be more than 3 minutes.

Retail Value of Automobiles

An authority claimed that the average retail sales price (without accessories) of four-cylinder two-door sedans built in the United States in 1969 was $2280.

A random sample of 100 is taken from sales records and its mean turns out to be $2330 with a standard deviation of $250. Does this suggest that the authority's claim is wrong? We want the likelihood of a Type I error to be 2%. (This, of course, makes $\alpha = 0.02$.)

$$H_0: \mu_x = \$2280$$
$$H_1: \mu_x \neq \$2280 \text{ (making the test two-sided)}$$
$$n = 100$$
$$s = \$250$$
$$\hat{\sigma}_{\bar{x}} = \frac{s}{\sqrt{n}} = \frac{250}{\sqrt{100}} = \frac{250}{10} = \$25$$
$$\alpha = 0.02$$
$$z_{\alpha/2} = 2.33$$
$$z_{\alpha/2}\hat{\sigma}_{\bar{x}} = 2.33(25) = 58.25 = \$58$$

We are assuming the original data precise to the nearest dollar.

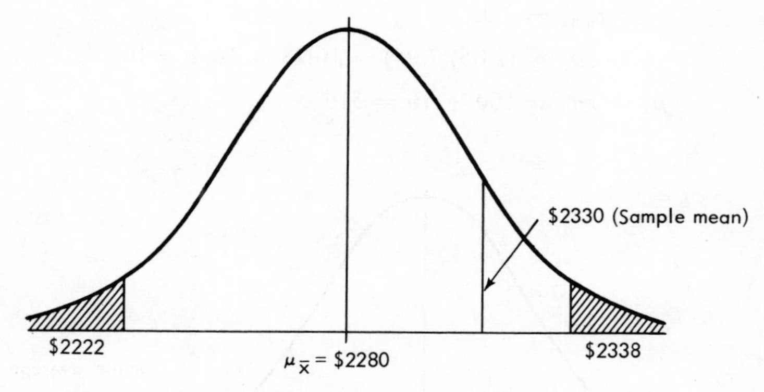

Distribution of sample means in dollars ($n = 100$, $\sigma_{\bar{x}} = \$25$)

The darkened areas are the rejection areas (rejection, that is, for the null hypothesis). The sample, as shown, is not so far from the mean as is the rejection area. So we do not reject the null hypothesis. This means that we do not reject the authority's claim that the average retail value of automobiles built in 1969 was $2280. This does not mean that we have proved the claim; we simply do not find sufficient grounds to reject it at a 0.02 level of significance.

Do not reject H_0. Accept H_1: the average car value is $2280.

Scholastic Aptitude Tests

Suppose that scholastic aptitude tests have a theoretical mean score of 500 with a standard deviation of 100. In a certain high school 160 randomly selected seniors took the test and averaged 524. Does this suggest, at a 0.02 level of significance, the school seniors are above average?

$$H_0: \mu = 500$$
$$H_1: \mu > 500$$
$$n = 160$$
$$\sigma = 100$$
$$\sigma_{\bar{x}} = \frac{\sigma}{\sqrt{n}} = \frac{100}{\sqrt{160}} = \frac{100}{12.6} = 7.94$$
$$\alpha = 0.02$$
$$z_{0.02} = 2.05$$
$$z\sigma_{\bar{x}} = (2.05)(7.94) = 16.28 = 16.3 = 16$$
$$\mu_{\bar{x}} + z\sigma_{\bar{x}} = 500 + 16 = 516$$

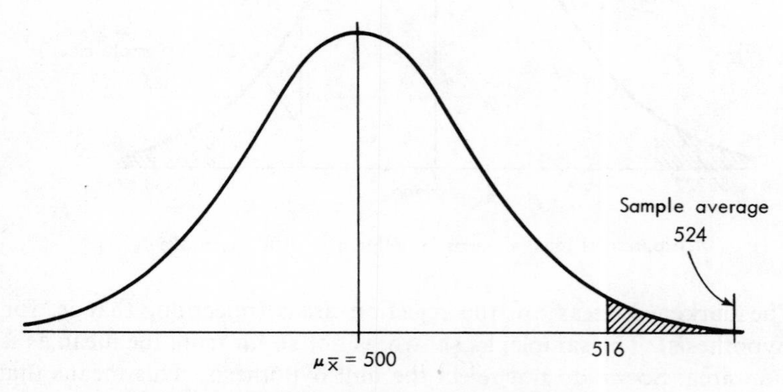

Distribution of sample means ($n = 160$, $\sigma_{\bar{x}} = 7.94$)

The sample is farther out than the decision line, so reject the null hypothesis that the high school is not significantly different from the average. This doesn't prove that it *is* significantly different, because the sample mean could, with $\mu_{\bar{x}} = 500$ and $n = 160$, fall beyond the 516 line 2% of the time just by chance. It says that the high-school scoring is probably above average, and it states *how* probable this is. The calculated risk of being wrong is 2%, a Type I error.

> Reject H_0. Accept H_1: the high-school scoring is significantly above average.

13-3 TYPE I AND TYPE II ERRORS, AGAIN

When you draw decision lines, you are saying that you will reject the null hypothesis if the sample mean falls in the rejection area beyond these lines (farther from the mean than these lines). But even if the null hypothesis were true, the sample mean *could* fall in one of those areas; it is simply very improbable that it will. Thus you are running the risk of rejecting a true hypothesis, but you have specified that risk (the risk of a Type I error) by stating a level of significance.

What, then, about a Type II error? A Type II error is the accepting of a false hypothesis. Perhaps these sketches will refresh your memory.

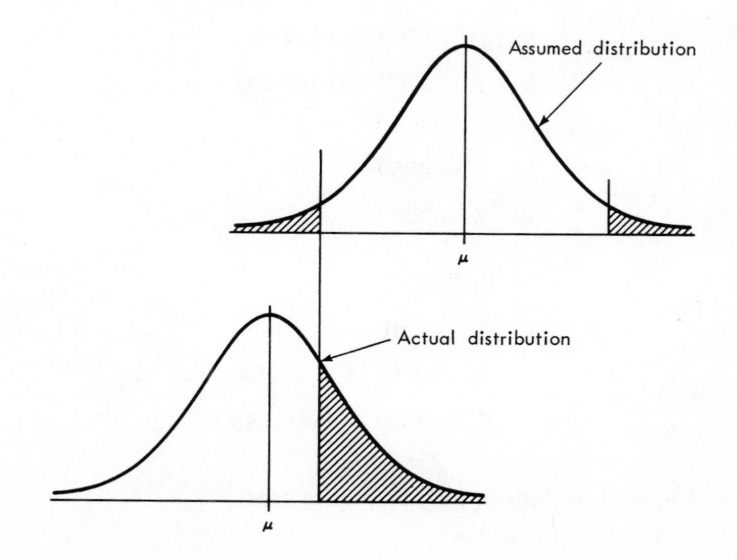

Using the same decision line as before, see how much of the time you will be accepting the assumed distribution when the actual distribution is as shown. The shaded space in the actual distribution shows the likelihood of a Type II error.

In the sampling procedures of quality control, the likelihood of making a Type I error, α (alpha)—the likelihood of rejecting a good lot of merchandise—is called the producer's risk. The likelihood of making a Type II error, β (beta)—the likelihood of accepting a poor lot—is called the consumer's risk.

A note of warning: A Type II error is accepting an hypothesis that is not true. But you cannot *calculate* this risk unless you know the μ and σ of the supposedly unknown population—an unlikely situation. On the other hand, maybe you would like to make a supposition about the actual population and, using your established decision plan, calculate the probability of a Type II error.

Gas Station

A real estate agent is eager to sell land for a gas station, claiming that, on the average, 5000 cars per day pass that property. A prospective customer suspects that the real estate agent's claim is too high, so he takes a complete count on 36 randomly selected days. He gets a mean of 4850 cars with a standard deviation of 500 cars. Is this enough less than 5000 to justify rejecting the agent's claim at a 1% level of significance?

$$H_0: \mu = 5000 \text{ cars per day}$$

$$H_1: \mu < 5000 \text{ cars per day}$$

$$n = 36$$

$$\bar{x} = 4850$$

$$s = 500$$

$$\hat{\sigma}_{\bar{x}} = \frac{s}{\sqrt{n}} = \frac{500}{6} = 83.3$$

$$\alpha = 0.01$$

$$z\hat{\sigma}_{\bar{x}} = 2.33(83.3) = 194 \text{ cars}$$

$$\mu - z\hat{\sigma}_{\bar{x}} = 5000 - 194 = 4806$$

The sample mean falls in the acceptance area.

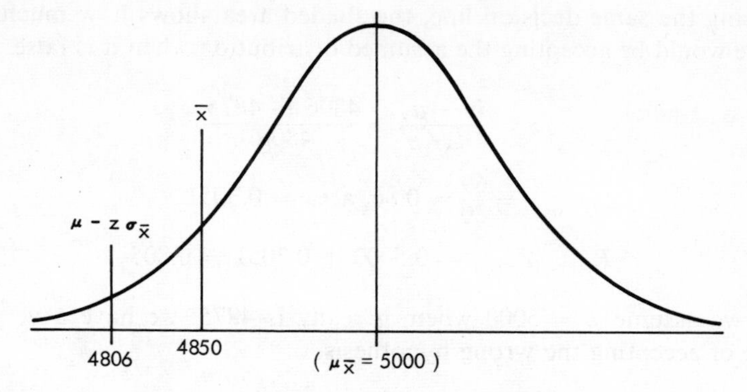

Distribution of sample means ($n = 36$, $\sigma_{\bar{x}} = 83.3$, $\alpha = 0.01$)

Do not reject H_0: the average of 5000 cars per day appears to be true.

Now ask yourself: Suppose I use this test and the mean number of cars per day is really 4875 with $\sigma = 480$. What would be the probability of accepting the 5000-car-per-day hypothesis? Again, what then would be the probability of a Type II error?

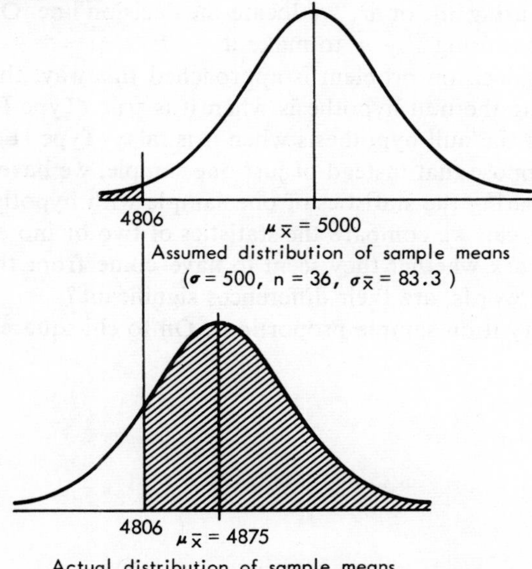

Assumed distribution of sample means
($\sigma = 500$, $n = 36$, $\sigma_{\bar{x}} = 83.3$)

Actual distribution of sample means
($\sigma = 480$, $n = 36$, $\sigma_{\bar{x}} = 80.0$)

Using the same decision line, the shaded area shows how much of the time we would be accepting the assumed distribution when it is false.

$$z = \frac{\bar{x} - \mu_{\bar{x}}}{\sigma_{\bar{x}}/\sqrt{n}} = \frac{4806 - 4875}{480/6}$$

$$= \frac{69}{80} = 0.86, \text{ area} = 0.3051$$

$$P_{(\text{Type II error})} = 0.5000 + 0.3051 = 0.805$$

If we assume $\mu = 5000$ when μ really is 4875, we have a very good chance of accepting the wrong hypothesis.

13-4 SUMMARY

We start with an hypothesis about a population mean (length in centimeters, weight in grams, volume in liters). This is a null hypothesis, H_0, the hypothesis which, before we start, we have no reason to reject—our reasons for rejecting are null. If our sampling turns up sufficient reason to reject H_0 statistically, then we have to accept our predetermined alternative hypothesis.

Accepting or rejecting hypotheses is again a matter of where we draw the line on our distribution. If $n \geq 30$, we can assume the distribution of sample means to be approximately normal and use z. If $n < 30$, we use a t distribution. By using $z\hat{\sigma}_{\bar{x}}$ or $t\hat{\sigma}_{\bar{x}}$ we locate the decision line. Of course, $\hat{\sigma}_{\bar{x}}$ is an approximation, using $s/\sqrt{n}$ to make it.

Whenever a decision problem is approached this way, there is the risk of either rejecting the null hypothesis when it is true (Type I error) or the risk of accepting the null hypothesis when it is false (Type II error).

But now suppose that instead of just one sample, we have two or more. Instead of comparing the statistics of one sample with hypothetical population parameters, can we compare the statistics of two or more samples with each other, and ask whether they seem to have come from the same population? In other words, are their differences significant?

First we'll try it on sample proportions. On to chi-square!

• 14 •

Chi-Square

Oft expectation fails.

WILLIAM SHAKESPEARE
All's Well That Ends Well (1602)

Ah, what a dusty answer gets the soul
When hot for certainties in this our life!

GEORGE MEREDITH
Modern Love, L (1862)

OBJECTIVES

To decide, in binomial situations, and even when more than two different responses are possible, whether there is a significant difference in sample counts.

SYMBOLS AND FORMULAS FIRST USED IN THIS CHAPTER:

$\text{d.f.} = (r - 1)(c - 1)$	degrees of freedom
$\chi^2 = \sum \frac{(a - e)^2}{e}$	chi-square
$\chi^2 = \frac{(\lvert a - e \rvert - 0.5)^2}{e}$	when d.f. = 1

14–1 EXERCISES

1. A geneticist claims that fruit flies (the famous *Drosophila*) appear white-eyed 0.25 of the time, red-eyed 0.75 of the time. In a sample of 2000, the number of white-eyed flies turned out to be 535. At a 0.01 level of significance, would you reject the geneticist's claim?

2. Problem 1 may be done by using chi-square or by showing a normal distribution of sample proportions. Now do it the way you didn't before. Do your conclusions agree?

3. A psychologist claimed that, among people 18 years old or more, a larger proportion of single people smoke than do married people. Random samples of 300 single and 300 married people were taken, revealing 160 single smokers and 140 married smokers. Considering the evidence given, what would you conclude about the psychologist's claim at a 0.01 level?

4. To test the effectiveness of a new pain-relieving drug, 80 patients in pain at a clinic are given a pill containing the drug while 80 others in pain are given a placebo (dummy pill). What can one conclude about the effectiveness of the drug if 56 of the patients in the first group felt a beneficial effect, and 38 of those who received the placebo felt a beneficial effect? Use $\alpha = 0.01$.

5. In a marketing study it was found that among 300 randomly selected housewives in Baltimore, 172 preferred Detergent A to Detergent B, while among 300 randomly selected housewives in Chicago, 203 preferred Detergent A to Detergent B. Use the level of significance $\alpha = 0.01$ to test the null hypothesis that there is probably no difference between the population proportions.

6. The business manager of the student council, in trying to decide which of two kinds of soft-drink vending machines to install in the student lounge, tests the machines by using each machine 250 times. The first kind of machine failed to work (that is, either failed to deliver a drink or made the wrong change) 19 times and the second kind of machine failed to work 8 times. Is this evidence (at the 0.05 level of significance) that the two machines are not equally good?

7. A sample survey of the weekly food expenditures for families with two children was conducted in two cities. In one city 100 families averaged

$38.25 with a standard deviation of $7.50, while 100 familes in the other city averaged $34.67 with a standard deviation of $8.15. To test whether there is a significant difference in sample *means*, you'll need the *analysis of variance*, not discussed until the next chapter.

8.

	Hair Color	
Eye Color	**Light**	**Dark**
Blue	32	12
Brown	14	22
Other	6	9

Are these characteristics independent? Or do eye color and hair color show more of a relationship than you would expect by chance alone? (Use a 0.05 level of significance.)

9.

20	70	10	5
80	50	50	20

Whatever the rows and columns are counts of, is there a significant difference in column proportions at a 0.05 level?

10. A psychologist wants to determine whether there is any relationship between height and self control. He gets the following results:

	Strong	**Average**	**Weak**
Under 5′8″	24	90	49
5′8″ to 6′1″	33	106	45
Over 6′1″	25	82	46

Use the level of significance $\alpha = 0.05$ to test the null hypothesis that there is no relationship.

11. The marketing research group of a particular firm conducted a survey in Buffalo, Kansas City, and San Diego to compare the sales potential of a new soft drink. Each person contacted was asked to try the new drink and classify it as excellent, good, or not satisfying. At a 0.01 level of significance, what do these results suggest?

Rating	**Cities** 1	2	3
Excellent	62	51	45
Good	28	30	35
Not satisfying	10	19	20

12. DS2, Aqual, and 707 are three drugs used in the cure of malaria. The number of cured patients, as well as the number of patients who again showed malarial symptoms, is shown below:

	DS2	**Aqual**	**707**	**Totals**
Number of cured cases	80	90	130	300
Number not cured	32	27	26	85
Totals	112	117	156	385

Is there sufficient evidence, at the 1% level of significance, to decide that the three drugs are not equally effective?

13. A textbook publisher asked faculty members for information on complimentary copies they had been supplied. The publisher tabulated responses according to academic degrees.

	Bachelor's	**Master's**	**Doctor's**
Commented	78	44	18
Did not respond	22	6	7

Do the number of responses indicate, at a 0.05 level of significance, that they are independent of the faculty member's degree?

14. Does this information suggest that at $\alpha = 0.01$ there is a relationship between income and the number of children in a family?

Family Income	**Number of Children** 0	1	2	More than 2
Under $10,000	14	26	49	43
$10,000 to $20,000	24	35	11	7
Over $20,000	7	12	8	9

15. An energetic psychology student was interested in the relative importance of certain factors in the selection of a female mate. He asked each in a random sample of 100 married men selected from different occupational groups in a particular industrial organization to check rating cards (very important, moderately important, not very important) on the importance of face, body, disposition, and intelligence. The cards were unsigned, but still the responses probably weren't especially reliable. However, opinion polls are usually much less reliable than the mathematics that goes with them.

	Face	**Body**	**Disposition**	**Intelligence**
Very important	15	18	63	20
Moderately important	62	61	19	74
Not very important	23	21	18	6

Determine what to test and then test it at a 0.05 level of significance.

14–2 REVIEW

We are going to take a backward glance at what it is we are about to go beyond.

Confidence Intervals

First, we shall estimate the proportion of hearts in a pack of playing cards; our already knowing the population proportion of hearts to be 0.250 gives a sort of Godlike opportunity to see how close chance will come to what we know to be true.

We draw a card, record it, replace it, and shuffle. If it were not for this replacing of the card each time, we could not use the binomial distribution because π, which is supposed to be constant, would change with every draw. Replacing the card each time keeps π constantly the same, as if we were drawing from an infinite population.

We get 10 hearts and 20 nonhearts. This gives us a p of $\frac{10}{30}$ or 0.333, which we have to use as π. Then np will be $(30)(\frac{10}{30}) = 10$ and $n(1 - p)$ will be $30(\frac{20}{30}) = 20$. As both are more than 5, the normal distribution should be a satisfactory approximation.

$$n = 30$$

$$p = \frac{10}{30} = 0.333$$

Again, using p for π,

$$\sigma_p = \sqrt{\frac{p(1-p)}{n}} = \sqrt{\frac{(0.333)(0.667)}{30}}$$

$$= \sqrt{\frac{0.2221}{30}} = \sqrt{0.00740}$$

$$= 0.0860$$

We decide on a 95% confidence interval, so

$$z = 1.96$$

$$p \pm z\sigma_p = 0.333 \pm 1.96(0.0860)$$

$$= 0.333 \pm 0.169$$

$$= 0.164 \text{ to } 0.502$$

This interval does include the known 0.250 for hearts. This was good exercise in confidence interval estimating, but it would have been easier to count the hearts in the whole pack—and it would have been exactly right. But in circumstances other than games of chance, it is usually impossible, or at least impractical, to count the whole population; so we accept this interval concept of estimation.

Testing Hypotheses

Now we might go on to this question: does our sample $p = (0.333)$ give us grounds for rejecting a null hypothesis that $\pi = 0.250$? What do you say intuitively?

Let's check it statistically.

$$H_0\colon \pi = 0.250$$

$$H_1\colon \pi \neq 0.250$$

$$\alpha = 0.05$$

$$n = 30$$

$$p = 0.333$$

$$\sigma_p = \sqrt{\frac{\pi(1-\pi)}{n}} = \sqrt{\frac{(0.250)(0.750)}{30}}$$

$$= \sqrt{\frac{0.1875}{30}} = \sqrt{0.00625} = 0.0791$$

$$\bar{p} \pm z\sigma_p = 0.250 \pm (1.96)(0.0791)$$

$$= 0.250 \pm 0.155$$

$$= 0.095 \text{ to } 0.405$$

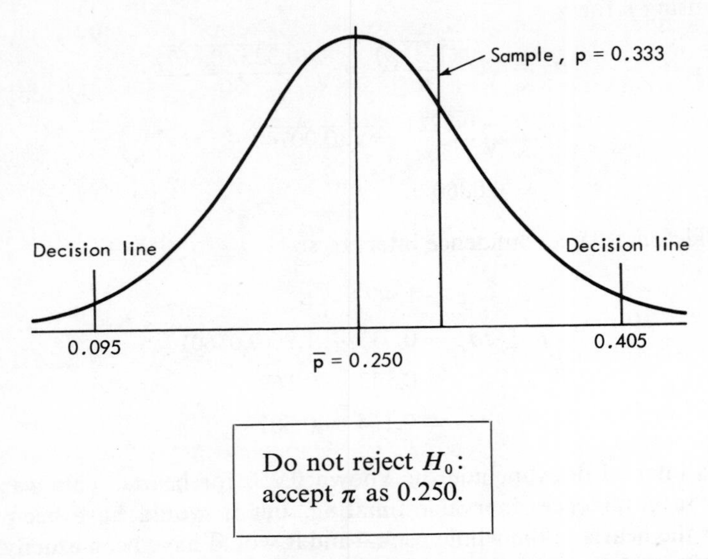

Do not reject H_0: accept π as 0.250.

This acceptance is in accordance with our Godlike knowledge of the facts. But life is not a small pack of cards.

14–3 CHI-SQUARE, TESTING HYPOTHESES

The techniques just reviewed in both estimation and hypothesis testing dealt directly with an assumed normal distribution. We shall now investigate the famous chi-square distribution.

Chi-square is pronounced *kī* (as in kite) and is written χ^2, which is the Greek for our *ch* and is formed by one curved and one straight line. We are going to show that in testing hypotheses about differences in sample proportions the chi-square table furnishes us with a more versatile technique than the *z* table we have just been reviewing.

Let us continue using our trivial example of drawing cards and recording the proportion of hearts. $H_0: \pi = 0.250$. $H_1: \pi \neq 0.250$.

Hearts	10
Nonhearts	20
Total	30

In this table there are two rows and one column. Call it a 2×1 table. When there are two rows and two columns, it is a 2×2 table. The intersections of rows and columns are called cells. In the 2×2 table there are four cells; in our 2×1 table there are two cells. In a 2×3 table there will be six cells. The number of cells is equal to the number of rows multiplied by the number of columns.

Now, what we intend to do is compare a maximum theoretical chi-square which we could expect by chance (call it χ_e^2) with the actual chi-square from our sample information (call it χ_a^2). Only if the actual χ_a^2 exceeds the maximum expected χ_e^2 shall we say that the differences are significant.

H_0: sample differences *not* significant ($\pi = 0.25$)

H_1: sample differences *are* significant ($\pi \neq 0.25$)
(implying, of course, that the samples come from different populations, not from ordinary playing cards)

First we shall go after χ_e^2. See Table V in the appendix. In order to get a value for the expected chi-square, we need two indicators: the level of significance and the so-called degrees of freedom. We select a level of significance, say $\alpha = 0.05$; the degrees of freedom we find from our table by seeing, given the row and column totals, how many counts we are free to choose. In this case, where the sample is 30 and where we get only 10 hearts, there must be 20 nonhearts. So the number of hearts is our only choice; that is, just 1 degree of freedom. Getting a certain number of hearts predetermines the number of nonhearts that must go with it. In our Table V on page 239 in the appendix, take the row for 1 under d.f. and the column for $\chi_{0.05}^2$ and arrive at 3.841. This is as large a χ_e^2 as we can expect by chance at $\alpha = 0.05$, if the samples come from the same or equal populations.

$$\chi_e^2 = 3.841$$

Now construct the 2×1 table, putting the expected counts in parentheses beside the actual counts. Then total the expected counts and expected proportions. The expected number of hearts is $0.250 \times 30 = 7.5$; the expected number of nonhearts must then be 22.5, according to H_0.

		Expected Proportions
Hearts	10 (7.5)	0.250
Nonhearts	20 (22.5)	0.750
Totals	30 (30)	1.00

To calculate the chi-square, subtract each expected count from its actual count, square this difference, and then divide the result by the expected count. Add the decimal values of all these separate fractions.

Look carefully at the formula. It is much clearer than our wordy explanation. (The greater the differences between the actual and the expected counts, the larger χ_a^2 will be.)

$$\chi^2 = \sum \frac{(\text{actual number minus expected number})^2}{\text{expected number}}$$

$$\chi^2 = \sum \frac{(a - e)^2}{e}$$

$$\chi^2 = \frac{(10 - 7.5)^2}{7.5} + \frac{(20 - 22.5)^2}{22.5}$$

$$= \frac{2.5^2}{7.5} + \frac{(-2.5)^2}{22.5} = \frac{6.25}{7.5} + \frac{6.25}{22.5}$$

$$= 0.833 + 0.278 = 1.111$$

I am unhappy to have to bring up an exception right off. But the exception happens to be where we have only 1 degree of freedom. In this case, the differences between actual and expected counts become somewhat exaggerated. But rather than scrapping our χ^2 for 1 degree of freedom (which would mean not using it for 2×1, 1×2, or 2×2 tables), we apply what is called *Yates' correction for continuity*. This simply amounts to decreasing all the differences by 0.5. If the difference is 3, make it 2.5; if the difference is -3, make it -2.5; this really amounts to decreasing the *absolute value* of the differences by 0.5. (If you decrease -3 as it stands by 0.5, you would actually get -3.5; this is not what we want. We want 3, the *absolute value*, decreased by 0.5, so that you get 2.5.)

We shall indicate the absolute value by drawing a vertical line on each side of an indicated difference. See how much clearer the formula is than all these words. Study it.

$$\chi_a^2 = \sum \frac{(|a - e| - 0.5)^2}{e}$$

Remember that this is applicable only when we have but 1 degree of freedom. There was only 1 degree of freedom in our problem of 10 hearts in a sample of 30 playing cards.

Let's go back to it.

$$\chi_a^2 = \sum \frac{(|a - e| - 0.5)^2}{e} = \frac{(2.5 - 0.5)^2}{7.5} + \frac{(|-2.5| - 0.5)^2}{22.5}$$

$$= \frac{4}{7.5} + \frac{4}{22.5} = 0.53 + 0.18 = 0.71$$

Our actual chi-square is less than the expected:

$$0.71 < 3.841$$

Do not reject H_0: the difference between actual and expected heart counts is not significant.

And this agrees with the conclusion we reached before, using the z table on page 237 in the appendix.

Snowhite, Again

The claim was that 4 out of 5 women preferred Snowhite. (Don't men brush their teeth, too?) H_0: there is no significant difference between the actual counts and what can be expected if $\pi = 0.80$ is true. H_1: there is a significant difference between actual and expected counts.

You see that 4 out of 5 is 0.80. Then 0.80×100 is 80, and 0.20×100 is 20.

		Expected Proportions
Prefer Snowhite	68 (80)	0.80
Don't prefer Snowhite	32 (20)	0.20
Totals	100 (100)	1.00

Get the χ^2_e. The d.f. = 1 and α is 0.01. Our table gives us χ^2 for $\alpha = 0.01$ or $\alpha = 0.05$. Use the column for $\alpha = 0.01$. Use one degree of freedom because, after the one choice of 68, there's no further choice; the number who don't prefer Snowhite has to be 32. *Attention*: Whenever we have a 2×1, 1×2, or 2×2 table, there is only one degree of freedom. All other tables have higher d.f.'s. In this case with 1.d.f. and $\alpha = 0.01$, we get $\chi^2_e = 6.635$.

$$\chi^2_{(\text{actual})} = \sum \frac{(|a - e| - 0.5)^2}{e}$$

Again, $|a - e|$ means the absolute value of the difference and, in effect, saves us any struggle with plus or minus signs.

$$\chi^2_a = \frac{(|68 - 80| - 0.5)^2}{80} + \frac{(|32 - 20| - 0.5)^2}{20}$$

$$= \frac{(12 - 0.5)^2}{80} + \frac{(12 - 0.5)^2}{20}$$

$$= \frac{11.5^2}{80} + \frac{11.5^2}{20} = \frac{132.2}{80} + \frac{132.2}{20}$$

$= 1.65 + 6.61$ (you can already see that χ_a^2 is going to exceed the 6.635 that is χ_e^2)

> Reject H_0: the proportion of women who prefer Snowhite is not 0.80; there is a significant difference between the actual and the expected preferences.

This agrees with our previous testing in Chapter 12 in that it tells us that there is a significant difference in sample proportions. But if you look closely, you will see that it is not quite so informative; it doesn't tell us which way the difference lies.

14–4 CHI-SQUARE, COMPARING TWO SAMPLES

Snowhite, Still Again

Suppose we take two samples and, finding a difference between the two sample proportions, ask whether this difference can be attributed just to chance variation between samples from the same population or whether the difference is enough to suggest that something is at work making the populations we drew from different. For instance, does the difference between a sample taken in the United States and a sample taken in Canada suggest two different populations—or could we expect this variation just by chance in samples from the same population?

In the Snowhite problem, 68% of the women in the United States sample preferred Snowhite toothpaste to other toothpastes. In another sample from 100 ladies in Canada, 86% did. (Some toothpaste! You'd better start buying stock instead of toothpaste.)

Now we are no longer considering the manufacturer's claim of an 80% preference. We are simply checking the sample differences. We lump the samples and get 154 out of 200, or 0.770, the proportion of all the ladies from both samples who prefer Snowhite. This is a point estimate of the population proportion from which the samples came. The question is: Do the sample counts support this hypothesis?

H_0: differences not significant ($\pi = 0.80$)

H_1: differences significant ($\pi \neq 0.80$)
(the samples appear to come from different populations)

$$\text{d.f.} = 1, \qquad \alpha = 0.01$$

$$\chi_e^2 = 6.635$$

Here are the two samples:

	No. 1	**No. 2**	**Totals**	**Assumed Proportions**
Preferred Snowhite	68 (77)	86 (77)	154	$\frac{154}{200} = 0.77$
Did not prefer Snowhite	32 (23)	14 (23)	46	$\frac{46}{200} = 0.23$
Totals	100 (100)	100 (100)	200	$\frac{200}{200} = 1.00$

$$\chi^2_{\text{actual}} = \sum \frac{(a - e)^2}{e}$$

but since d.f. = 1,

$$\begin{aligned}
\chi^2_{\text{actual}} &= \sum \frac{(|a - e| - 0.5)^2}{e} \\
&= \frac{(|68 - 77| - 0.5)^2}{77} + \frac{(|86 - 77| - 0.5)^2}{77} \\
&\quad + \frac{(|32 - 23| - 0.5)^2}{23} + \frac{(|14 - 23| - 0.5)^2}{23} \\
&= \frac{8.5^2}{77} + \frac{8.5^2}{77} + \frac{8.5^2}{23} + \frac{8.5^2}{23} \\
&= \frac{72.2}{77} + \frac{72.2}{77} + \frac{72.2}{23} + \frac{72.2}{23} \\
&= 0.94 + 0.94 + 3.1 + 3.1 \\
&= 8.08
\end{aligned}$$

$$\chi^2_a > \chi^2_e, \qquad 8.08 > 6.635$$

> Reject H_0: the samples show significantly different preferences for Snowhite.

This was a 2×2 table; we first used a 2×1 table. But we can have tables with any number of rows and columns. This not only frees us from being restricted to a binomial situation, allowing us more than two categories, but also allows us as many column classifications as we want—United States, Canada, New Zealand, Fiji . . . .

Except in the 2×1 and 1×2 tables, you can get the degrees of freedom by multiplying the number of rows minus one by the number of columns minus one: d.f. $= (r - 1)(c - 1)$. When d.f. is more than 1, there is no need for the Yates' correction of -0.5; this, I am sure, you'll be glad to know.

$$2 \times 2 \text{ table:} \quad \text{d.f.} = (2 - 1)(2 - 1) = 1$$
$$2 \times 3 \text{ table:} \quad \text{d.f.} = (2 - 1)(3 - 1) = 2$$
$$3 \times 4 \text{ table:} \quad \text{d.f.} = (3 - 1)(4 - 1) = 6$$
$$6 \times 5 \text{ table:} \quad \text{d.f.} = (6 - 1)(5 - 1) = 20$$

Notice particularly that after you put as many expected counts into your table as there are degrees of freedom, the rest of the expected counts are predetermined by the totals. I shall call this to your attention again in a few minutes when it has the potential of saving you much time and labor.

14-5 BEYOND THE 2 × 2 TABLE

The greatest power lies ahead. The chi-square technique may be used for any number of rows or columns! Onward!

House Heating Fuels

The following table shows the results of asking 420 house owners to rate the heating fuel they were using.

	Oil	Gas	Electricity	Totals	Proportion
Excellent	40 (31)	60 (51)	20 (38)	120 (120)	$\frac{120}{420} = 0.286$
Unimpressive	50 (47)	50 (77)	80 (56)	180 (180)	$\frac{180}{420} = 0.429$
Poor	20 (32)	70 (52)	30 (36)	120 (120)	$\frac{120}{420} = 0.286$
Totals	110 (110)	180 (180)	130 (130)	420 (420)	$\frac{420}{420} = 1.01$*

*Not exactly 1.00 due to rounding off.

The number of degrees of freedom equals the number of rows minus one times the number of columns minus one: d.f. $= (r - 1)(c - 1) = (3 - 1)(3 - 1) = 2 \times 2 = 4$. Now, as I suggested for a timesaver, compute four expected counts. Multiply the column totals by the row proportions. For instance, $0.286(110) = 31$; $0.286(180) = 51$; $0.428(110) = 47$; $0.428(180) = 77$. All the rest of the cell counts (because you have just four degrees of freedom) will be determined by the four you have computed and by the totals.

I have circled the expected counts that you computed; the others are then predetermined by the row and column totals. When there are four degrees of freedom like this, you need to compute only four expected cell counts—but not *all* the expected cell counts in any row or column because by computing one that is predetermined already, you waste a degree of freedom.

We shall hypothesize that there is not a significant difference between actual counts. The alternative hypothesis is, of course, that there *is* a significant difference in sample counts.

H_0: there is no significant difference

H_1: there is a significant difference

$$\alpha = 0.01$$

$$\chi_e^2 = 13.277$$

Are we going to exceed 13.277?

$$\chi_a^2 = \sum \frac{(a - e)^2}{e} \qquad \text{(no correction for continuity necessary)}$$

$$= \frac{(40 - 31)^2}{31} + \frac{(60 - 51)^2}{51} + \frac{(20 - 38)^2}{38}$$

$$+ \frac{(50 - 47)^2}{47} + \frac{(50 - 77)^2}{77} + \frac{(80 - 56)^2}{56}$$

$$+ \frac{(20 - 32)^2}{32} + \frac{(70 - 52)^2}{52} + \frac{(30 - 36)^2}{36}$$

$$= \frac{9^2}{31} + \frac{9^2}{51} + \frac{(-18)^2}{38} + \frac{3^2}{47} + \frac{(-27)^2}{77} + \frac{24^2}{56}$$

$$+ \frac{(-12)^2}{32} + \frac{18^2}{52} + \frac{-6^2}{36}$$

$$\left.\begin{aligned} &= \frac{81}{31} + \frac{81}{51} + \frac{324}{38} + \frac{9}{47} + \frac{729}{77} + \cdots \\ &= 2.6 + 1.6 + 8.5 + 0.2 + 9.5 + \cdots \end{aligned}\right\} \text{do these together}$$

Whoa! The sum 22.4 is more than 13.277 already!

Already we exceed the expected $\chi_e^2 = 13.277$. So, at a 0.01 level, we reject the hypothesis that the differences in sample counts are due to chance alone. We have to accept the hypothesis that there is some significance to the differences. There *is* a relationship between these different furnace fuels and the customer's satisfaction with them. But that's all; nothing tells which fuels show such a relationship nor what that relationship is numerically.

Shampoos (from Hindi, capo)

Chi-square, as you are beginning to see and will certainly see if you do the exercises, is a versatile and revealing approach to information on counts.

Our last example is a 3 × 4 table. Looking at this example in different ways may bring you closer to a real understanding.

Suppose we have simply the totals of rows and columns.

					Totals
					135
					82
					40
Totals	48	77	71	61	257

As shown, we could next put each row total over the grand total. Lumping the information this way, we can use these proportions of the column totals for expected counts in the respective cells of that column. The assumption is that the differences among actual and expected counts are not significant. This is the null hypothesis and it justifies putting the row sum over the grand total to estimate the population proportions.

Before we go on, and before we know any more about the problem, we can get χ_e^2. As $\alpha = 0.05$, and d.f. $= (r - 1)(c - 1) = (3 - 1)(4 - 1) = (2)(3) = 6$, then $\chi_e^2 = 12.592$. Whatever figures appear in the problem, this is the figure that is the ultimate measure of significant difference.

Now let's look at our expected figures. We calculated six cells. There are

					Totals	**Proportions**
	(0.525)(48)=25	(0.525)(77)=40	(0.525)(71)=37		135	$\frac{135}{257}=0.525$
	(0.319)(48)=15	(0.319)(77)=25	(0.319)(71)=23		82	$\frac{82}{257}=0.319$
					40	$\frac{40}{257}=0.156$
Totals	48	77	71	61	257	$\frac{257}{257}=1.000$

six degrees of freedom. The other counts must be predetermined by the totals.

					Totals
	25	40	37	32	134
	15	25	23	19	82
	8	12	11	10	41
Totals	48	77	71	61	257

We have rounded off our expected values to the nearest unit. Considering the assumptions made and the approximations involved, carrying the expected frequencies to one decimal place would be unreasonable. However, if you round off the calculated counts, some totals may vary slightly from the originals. Perhaps you will want to make some reasonable adjustments so that the sums do come out the same, but it's not essential.

Now plug in the missing information from an actual problem.

					Totals	**Proportions**
	15 ㉕	27 ㊵	50 ㊲	43 (32)	135	$\frac{135}{257} = 0.525$
	25 ⑮	37 ㉕	12 ㉓	8 (19)	82	$\frac{82}{257} = 0.319$
	8 (8)	13 (12)	9 (11)	10 (10)	40	$\frac{40}{257} = 0.156$
Totals	48 (48)	77 (77)	71 (71)	61 (61)	257	1.000

The startling fact is that we can proceed now, which we shall do, to calculate χ^2_a without knowing what the problem is all about. We state only what it is we are testing. This is it: there is no significant overall difference between the actual and the expected counts in the different cells.

There is some advantage in doing all the calculations for each cell as we go along; in this way we may detect early that we have already exceeded χ^2_e and don't have to continue calculation beyond that point.

Let's see.

$$\alpha = 0.05$$

$$\text{d.f.} = 6$$

$$\chi^2_e = 12.592$$

$$\chi_a^2 = \sum \frac{(a-e)^2}{e}$$

$$= \frac{(15-25)^2}{25} + \frac{(27-40)^2}{40} + \frac{(50-37)^2}{37} + \cdots$$

$$= \frac{(-10)^2}{25} + \frac{(-13)^2}{40} + \frac{13^2}{37} + \cdots$$

$$= \frac{100}{25} + \frac{169}{40} + \frac{169}{37} + \cdots$$

$$= 4.00 + 4.23 + 4.57 = 12.80$$

We have already exceeded 12.592!

So this means that the variation from expected counts *is* significant. It means that there is something about the columns that makes the row counts different from what we would expect just by chance. There is some relationship involved. We don't say what the relationship is or exactly where it is; we simply say that these variations can't be attributed to chance alone.

Reject H_0: the differences are significant.

Now these figures can apply to any categories (rows) and classifications (columns)—blueberry pickings in different areas, defects in structural steel from different companies, attractiveness of males (or females) from different colleges, cosmic ray readings from a satellite at different levels of the stratosphere.

We have already performed the necessary mechanics of the problem. Let's see now what it might mean if we return to our problem concerning different brands of shampoos and their customer ratings.

H_0: no significant difference in shampoo ratings

H_1: there *is* a significant difference in shampoo ratings

	Shampoo Brands				
Ratings	**A**	**B**	**C**	**D**	**Totals**
Above average	15	27	50	43	135
Average	25	37	12	8	82
Below average	8	13	9	10	40
Totals	48	77	71	61	257

$$\chi_a^2 = 12.80^+ > \chi_e^2 = 12.592$$

Reject H_0: there *is* a significant difference in shampoo ratings for the different brands.

14-6 SUMMARY

Chi-square is a basic statistical concept. It furnishes us with a technique for deciding whether the differences between population proportions and the proportions of a sample are significant, whether the difference between the two sample proportions is significant, or whether the differences among the proportions in more than two samples are significant. This gives us great versatility.

The level of significance tells us how much of the time we will be rejecting a true null hypothesis just by chance. As usual, some assumptions are made: the distribution of counts about the mean count is assumed to be normal; the standard deviation of these counts is assumed to be estimable from the samples. Though these assumptions are pretty well obscured, mention of them serves to remind us that our results are still approximations. If the χ_a^2 values come out uncomfortably close to the χ_e^2 value, the best decision is to withhold judgment and take another sample (maybe even a larger one).

In our attempts to cope with an uncertain world, we accept these approximations because they are helpful. In a hospital situation the directions may be: if the patient's electrocardiogram registers more than a 2-millimeter depression, call the doctor at once. This involves several statistical approximations, but it is a lot better than saying, "Call the doctor if you think you ought to."

The next basic concept to be considered in our statistical drama is the *analysis of variance*, and, as you might expect, it has to do with means rather than proportions. Perhaps you remember that you couldn't do Problem 7 in the exercises for this chapter because, as yet, we have had no way of comparing different sample means. In the next chapter we shall do just that by using variances (unsquare-rooted standard deviations, if you want to look at it that way).

But for now, before the exercises, see Table VIII in the appendix for a summary of the chi-square routine.

• 15 •

The Analysis of Variance

Undoubtedly one of the most elegant, powerful, and useful techniques in modern statistical method is that of the Analysis of Variation . . . .

M. J. MORONEY
Facts from Figures (1956)
[Check your library.
This is a good one.]

OBJECTIVES

To decide whether the difference between sample means is more than we would expect by chance, given the hypothesis that the samples come from the same population.

SYMBOLS AND FORMULAS FIRST USED IN THIS CHAPTER:

n.d.f. $= c - 1$	numerator degrees of freedom
d.d.f. $= c(r - 1)$	denominator degrees of freedom
F_e	maximum expected ratio using d.f. and α
ANOVA	the analysis of variance

$$\hat{\sigma}^2_{\text{within}} = \frac{s_1^2 + s_2^2 + \cdots + s_c^2}{c}$$

$$\hat{\mu}_{\bar{x}} = \frac{\bar{x}_1 + \bar{x}_2 + \cdots + \bar{x}_c}{c}$$

$$\hat{\sigma}^2_{\text{between}} = r\left[\frac{\sum (\bar{x} - \mu_{\bar{x}})^2}{c - 1}\right]$$

$$F_a = \frac{\hat{\sigma}^2_b \ (b = \text{between})}{\hat{\sigma}^2_w \ (w = \text{within})}$$

15–1 EXERCISES

1. Mrs. Jones timed herself on two different routes to her sister's home. Here are the results in minutes:

Route 1	Route 2
40	30
41	32
30	34

The samples are too small to be very reliable, but they keep the arithmetic simple. (And, of course, we would want to know whether they were taken at the same times of day, the same days of the week, the same part of the year, under the same weather conditions, etc.)

Now check your null hypothesis, using the analysis of variance at $\alpha = 0.05$. Don't neglect to formulate an alternative hypothesis.

2. Two brands of surgical silk were tested for tensile strength in pounds per square inch, yielding the following results:

Brand A	Brand B
64	82
74	77
66	84
82	72
79	76

At a 0.05 level of significance, do the samples indicate a difference in the means of the two brands?

3. Four lots of 5 pigs were randomly selected. Each lot, for one year after suckling, was fed a different commercial feed exclusively, with the following results in pounds gained:

Feed A	Feed B	Feed C	Feed D
$\bar{x}_A = 141$	$\bar{x}_B = 133$	$\bar{x}_C = 183$	$\bar{x}_D = 190$
$s_A^2 = 45$	$s_B^2 = 556$	$s_C^2 = 311$	$s_D^2 = 812$

At a 0.01 level of significance does this indicate a difference in the effectiveness of the feeds?

Of course, even if the pigs ate the same amounts of food every time (which the exercise statement doesn't insist on), there might be metabolic differences in the pigs themselves which could somewhat account for the differences.

4. Four different milling machine operators turned out the following number of pieces on the same model of machine in five consecutive hours. At a 0.05 level of significance do the results indicate a difference in operator performance?

Abbott	Babbage	Cojari	D'Alembert
50	60	70	80
60	85	60	55
70	85	50	65
65	40	45	70
65	50	65	60
$\bar{x}_A = 62$	$\bar{x}_B = 64$	$\bar{x}_C = 58$	$\bar{x}_D = 66$
$s_A^2 = 58$	$s_B^2 = 418$	$s_C^2 = 108$	$s_D^2 = 92$

Where, in your calculations, do you decide not to go any further? Why?

5. A hostility test is designed by a psychologist who claims that it measures a person's aggressive resentment at the situations he finds himself in. He takes 10 people at random from each of the two groups which he considers to be "campus dissenters" and "college faithfuls." He gives them his test and gets a mean of 78 for the first group and 63 for the second, and variances of 5 and 2, respectively. At $\alpha = 0.01$ show whether this test differentiates between the two groups.

6. Four different makes of cars of the same model, year, and approximately the same mileage were checked for miles per gallon with the following mean results:

	Make of Cars			
	1	**2**	**3**	**4**
	21	18	15	13
Gas Mileage	20	18	16	13
	19	18	14	15

Using the analysis of variance what null hypothesis would you test?

Test it at $\alpha = 0.05$. (Do you realize that when you use $\hat{\mu}_{\bar{x}}$ in this case, you are probably using the mean of the means of the means?)

7. Now change the situation in the above problem to car counts.

Brands of Gas	Make of Cars 1	2	3	4
A	21	18	15	13
B	20	18	16	13
C	19	18	14	15

What hypothesis would you test this time? In the first situation you wanted to test whether there was a significant difference between mean gas mileages for the four makes of cars. The second situation is complicated by bringing in different brands of gasoline. This will force you to use chi-square instead of the analysis of variance. Think it over carefully and record your hypothesis before you tangle with the arithmetic. (Thinking ahead is a *very* important habit to develop.)

8. A nationally known guitarist claims that 80% of all his recordings are bought by blacks. A random sample of 100 buyers shows 35 were bought by whites. If you are willing to reject a true hypothesis no more than 5 times in 100, what would this sample suggest about the guitarist's claim?

9. The average waist of men over forty years old is said to measure 38 inches with a standard deviation of 2 inches. What is the probability that an average of 36 inch waists from a random sample of 100 men will fall in a $\mu \pm 2.33\sigma_{\bar{x}}$ interval?

10. Four furniture enamel brands were tested and given an index of durability in each of three tests. Do the following indices indicate a significant difference in enamel brand durability at = 0.01?

Paint Brands A	B	C	D
17	12	23	13
15	15	17	17
22	9	20	12

11. The superintendent of a supermarket specifies the number of cans of shelf frontage to be allowed different brands of different products. In

an effort to check whether the frontage makes any significant difference, he accumulated this information on the number of cans of clam chowder sold in a week.

Shelf Frontage in Number of Cans	Brand A	Brand B	Brand C
3	43	44	59
4	53	54	45
5	56	50	56

Select a level of significance. Clearly interpret your testing.

15–2 THE ANALYSIS OF VARIANCE—ANOVA

With chi-square we tested hypotheses that samples came from the same population (or populations with the same proportions)—that differences in sample proportions were due to chance only and therefore were not significant. With the analysis of variance, we are going to test hypotheses that samples come from the same population (or populations with the same means)—that the differences in sample means are due to chance only and therefore are not significant.

In chi-square analysis we use counts: white-eyed fruit flies, single people who smoke, patients who show a beneficial effect from a specific drug. These counts we put in the cells of our intersecting rows and columns. Now we are going to use measurements from any number of samples of any sample size, and we will put these measurements in separate columns for each sample.

Our key ratio involves variances, which are simply standard deviations before taking the square roots. When we found standard deviations, we took all the deviations from the mean, squared them, got their average, and then took the square root. Now we won't take the square root at all; the variance will serve us just as well. We estimate the population variance by using the variance *between* the sample means; then we estimate the population variance by using the variances *within* the samples. Putting the first over the second will give us the F ratio. (The F is for Sir Ronald A. Fisher, who did so much original work in this area.)

We shall check this actual F ratio against the maximum F ratio we might expect by chance with samples from the same population—at a given level of significance. Just as we compared the actual χ_a^2 with the expected χ_e^2, so we shall compare the actual F_a with the expected F_e. The null hypothesis, again, will be that the variation in sample means is only what we could expect by chance from the given population. The alternative hypothesis will be that a relationship exists which is not just chance, that there are different populations.

We can't avoid getting involved in some fairly extensive calculations, but they are a means to an end. The important thing for you to recognize and appreciate is that a ratio made up of the estimated variance from between samples and the estimated variance from within samples may be used to detect differences that are beyond chance. The analysis of variance is a tremendously important concept in this respect. Perhaps you will remember more about it if you go through the agony and ecstasy of doing all the necessary calculations for a case or two—or three.

Stone Samples

Geologists sometimes use the Wentworth rock-size scale, with grade names of boulders, cobbles, pebbles, sand, silt, and clay. Random samples of 3 (impractically small, but the small sample size will save us a lot of calculation in our model problem) were taken from three different glacial cobble deposits and their lengths recorded in inches. Our question becomes: Are the mean sizes of these cobble samples what we could expect if they came from the same populations, or are they significantly different? Use $\alpha = 0.05$.

H_0: no significant difference between sample means

H_1: differences between sample means are significant (suggesting that the cobble deposits did not come from the same population)

Sample 1	Sample 2	Sample 3
1	2	6
3	2	6
5	8	3

From a statistical point of view, these sample measurements could be anything—a dimension in appropriate units from spermatazoa, mosquitoes, skyscrapers. The question remains the same: Is there a significant difference in sample means? And you can make a decision about this by performing the technical procedures involved in the analysis of variance.

15–3 THE EXPECTED VALUE OF F

The simplicity of this part of the analysis of variance is encouraging. Given c number of samples (columns) with r items in each sample (rows), it can be shown mathematically that, if the samples come from the same normal population or normal populations having the same σ, an F ratio for any given level of significance may be used. This theoretical ratio is dependent on the sample size, the number of samples, and the level of significance (and the assumption of a normal distribution).

Go to Table VI in the appendix (page 240). Table VI has two parts: (a) for $\alpha = 0.05$, (b) for $\alpha = 0.01$, a separate page for each level of significance. (If you want F for other levels of significance, try a volume of statistical tables at a sizable library.) Select the part of the table that goes with your level of significance, say $\alpha = 0.05$ in this cobble problem. You need to know that the numerator of the F ratio is calculated with $c - 1$ degrees of freedom and the denominator with $c(r - 1)$ degrees of freedom. In this case, the numerator degrees of freedom are $c - 1 = 3 - 1 = 2$, and the denominator degrees of freedom are $c(r - 1) = 3(3 - 1) = 3(2) = 6$. The numerator degrees of freedom head the vertical columns in Table VI(a) and Table VI(b); the denominator degrees of freedom lead the horizontal rows. At the intersection of the respective rows and columns you will find, in this case, $F_e = 5.14$ (use F sub e for the expected maximum F at α).

If our actual F exceeds this theoretical value, we are going to reject the null hypothesis that the difference in sample means is not significant and conclude that some sort of significant relationship exists.

15–4 THE ACTUAL VALUE OF F

The actual value of F is, as we have said, the ratio of estimated variance between samples over the estimated variance within samples. In short,

$$F_a = \frac{r\hat{\sigma}_{\bar{x}}^2}{\hat{\sigma}^2}$$

Our problem has three samples of cobble lengths in inches. First compute their means, then their variances (which are simply the average squared deviations from their means). See table at top of next page. At first we will be working for the estimated variance *within* samples, later for the estimated variance *between* them.

Sample #1	Sample #2	Sample #3
1	2	6
3	2	6
5	8	3
$\bar{x}_1 = \frac{\sum x}{r} = \frac{9}{3} = 3$	$\bar{x}_2 = \frac{\sum x}{r} = \frac{12}{3} = 4$	$\bar{x}_3 = \frac{\sum x}{r} = \frac{15}{3} = 5$

The sample size (use r instead of n here) will always be the number of rows in the sample; again, in our analysis of variance we are using r for n; for the *number* of samples we use c (columns); this conforms with our χ^2 technique.

$$s_1^2 = \frac{\sum (x - \bar{x}_1)^2}{r - 1} = \frac{(-2)^2 + 0^2 + 2^2}{2} = \frac{8}{2} = 4$$

$$s_2^2 = \frac{\sum (x - \bar{x}_2)^2}{r - 1} = \frac{(-2)^2 + (-2)^2 + 4^2}{2} = \frac{24}{2} = 12$$

$$s_3^2 = \frac{\sum (x - \bar{x}_3)^2}{r - 1} = \frac{1^2 + 1^2 + (-2)^2}{2} = \frac{6}{2} = 3$$

With this information, we can guess at the population variance by using the mean of the variances within the three samples.

$$\hat{\sigma}^2_{\text{within}} = \frac{s_1^2 + s_2^2 + s_3^2}{c} = \frac{4 + 12 + 3}{3} = \frac{19}{3} = 6.33$$

This is an estimated variance from *within* the samples and will be the denominator of our actual F ratio.

Now for the numerator. First a point estimate of the mean of the sample means.

$$\hat{\mu}_{\bar{x}} = \frac{\bar{x}_1 + \bar{x}_2 + \bar{x}_3}{c} = \frac{3 + 4 + 5}{3} = \frac{12}{3} = 4.0$$

Now calculate the variance of the individual sample means from this estimated population mean.

$$\hat{\sigma}^2_{\bar{x}} = \frac{\sum (\bar{x} - \hat{\mu})^2}{c - 1} = \frac{(3 - 4)^2 + (4 - 4)^2 + (5 - 4)^2}{2}$$

$$= \frac{(-1)^2 + 0^2 + 1^2}{2} + \frac{1 + 1}{2} = \frac{2}{2} = 1$$

In previous chapters when we were dealing with the distribution of *sample* means, we used

$$\hat{\sigma}_{\bar{x}} = \frac{s}{\sqrt{n}}$$

which is an approximation from

$$\sigma_{\bar{x}} = \frac{\sigma}{\sqrt{n}}$$

Now we are interested in the estimated population variance. A little algebra helps.

$$\hat{\sigma}_{\bar{x}} = \frac{\hat{\sigma}}{\sqrt{n}}$$

$$\sqrt{n}\,\hat{\sigma}_{\bar{x}} = \hat{\sigma}$$

$$n\hat{\sigma}^2_{\bar{x}} = \hat{\sigma}^2$$

$$\hat{\sigma}^2 = n\hat{\sigma}^2_{\bar{x}}$$

Since $n = r$, then

$$\hat{\sigma}^2 = r\hat{\sigma}^2_{\bar{x}}$$

So in the geology problem,

$$\hat{\sigma}^2 = r\hat{\sigma}^2_{\bar{x}} = 3(1) = 3$$

This is the estimated variance from *between* the samples and gives us the numerator of our actual F_a ratio. Use F sub a for the actual F.

$$F_a = \frac{\hat{\sigma}^2_{\text{(between)}}}{\hat{\sigma}^2_{\text{(within)}}} = \frac{3}{6.33} = 0.474$$

This F ratio is certainly below our expected $F_e = 5.14$, so do *not* reject the null hypothesis. We don't have evidence in the samples that will allow us to conclude that they did not come from the same population.

Do not reject H_0: the differences in cobble size are not significant.

A theoretical F distribution looks about like this:

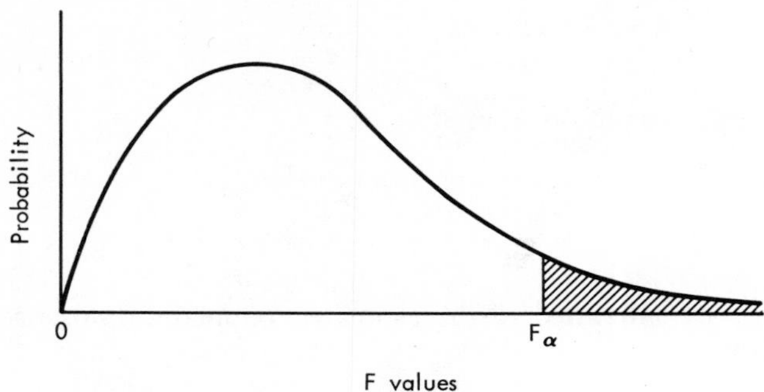

The F values range from zero upward. If the samples we are using come from the same population, then the F ratio formed by the estimated variances is not likely to fall above F_e. If it does exceed F_e, we will reject H_0.

This is the famous analysis of variance. We are comparing our actual F ratio of estimated variances with the largest F ratio of variances which we might expect by chance from a given population at a given level of significance. That's the important concept.

15–5 COMPUTATIONAL PROCEDURES

1. The expected value of F (from the table) depends on two factors: the level of significance and the degrees of freedom. There is a separate table for each level of significance. Degrees of freedom are determined by the number of samples and their sizes. In each table F is found where the numerator degrees of freedom (the vertical columns) and the denominator degrees of freedom (the horizonal rows) intersect. The numerator degrees of freedom are: n.d.f. $= c - 1$, where c is the number of columns (actually the number of samples). The denominator degrees of freedom are: d.d.f. $= c(r - 1)$, where r is the number of rows (actually the sample sizes).

Look up F_e first, as we did with χ_e^2, and later calculate the actual problem F.

$$\text{n.d.f.} =$$

$$\text{d.d.f.} =$$

$$\alpha =$$

$$F_\alpha =$$

2. Calculate the separate sample variances in order to estimate the population variance from within. First compute the separate sample means:

$$\bar{x} = \frac{\sum x}{r}$$

Then find each sample variance.

$$s^2 = \frac{\sum (x - \bar{x})^2}{r - 1}$$

Computing all these variances may be a good exercise in arithmetic. But after you've done it a couple of times, we're going to put the sample variances right in the exercise, saving you some spiritual wear and tear.

The population variance estimated from the variances within the samples is then found by taking an average of these variances. (The hat, remember, emphasizes "estimated.")

$$\hat{\sigma}^2_{\text{(within)}} = \frac{s_1^2 + s_2^2 + s_3^2 + \cdots + s_c^2}{c}$$

3. Estimate the population variance from between samples. First estimate the population mean. (Yes, this really is the mean of the means again.) This is an estimate but it's all we have. A practical lesson that keeps appearing in this course is that, if you haven't got what you need, find something that you can, with some confidence, use in its place.

$$\hat{\mu}_{\bar{x}} = \frac{\bar{x}_1 + \bar{x}_2 + \bar{x}_3 + \cdots + \bar{x}_c}{c}$$

Then

$$\hat{\sigma}^2_{\bar{x}} = \frac{\sum (\bar{x} - \hat{\mu}_{\bar{x}})^2}{c - 1}$$

That $c - 1$ is our old $n - 1$. Why? Because the number of means we are finding the variance of is the number of samples, which is the same as the number of columns.

Notice that we have just estimated $\sigma^2_{\bar{x}}$, not σ^2. Remember that two pages back we got σ^2 in the following manner:

$$\hat{\sigma}^2_{\text{(between)}} = r\hat{\sigma}^2_{\bar{x}}$$

4. Actual F ratio: Having now the variance of the population estimated from between the samples and the variance of the population estimated from within the samples, we can get our actual F.

$$F_a = \frac{\hat{\sigma}^2_{\text{(between)}}}{\hat{\sigma}^2_{\text{(within)}}}$$

5. Comparison: If the actual F comes out less than the maximum expected F_e, then do not reject the null hypothesis that there is no significant difference in sample means. Otherwise accept the alternative hypothesis that there *is* a significant difference in means.

Outline

To put it all together once more:

ANOVA (Analysis of Variance)

1. n.d.f. $= c - 1 =$

   d.d.f. $= c(r - 1) =$

   $\alpha =$

   Look up F_e (or call it F_α)

2(a). $\bar{x} = \dfrac{\sum x}{c}$

$$s^2 = \frac{\sum (x - \bar{x})^2}{r - 1}$$

2(b). $\hat{\sigma}^2_{(\text{from within})} = \dfrac{s_1^2 + s_2^2 + s_3^2 + \cdots + s_c^2}{c}$

3(a). $\hat{\mu}_{\bar{x}} = \dfrac{\bar{x}_1 + \bar{x}_2 + \bar{x}_3 + \cdots + \bar{x}_c}{c}$

3(b). $\hat{\sigma}^2_{(\text{from between})} = r\left[\dfrac{\sum (\bar{x} - \hat{\mu}_{\bar{x}})^2}{c - 1}\right]$

4. $F = \dfrac{\hat{\sigma}_b^2}{\hat{\sigma}_w^2}$

5. How does this ratio compare with F_e? So what about the null hypothesis that there is no difference between the populations from which the samples came (that there is no significant difference between sample means)?

The analysis of variance is another expedition in hypothesis testing. We are trying to determine whether the differences in mean measurements between two samples or among several samples are significant, suggesting that the samples came from different populations, suggesting that there seems to be some relationship between samples and the sources from which they came—differences in times of different routes to work, differences in tensile strength of surgical silks, differences in nutrition values of pig feeds, productivity of milling machine operators, mileage for gasoline brands. In all cases we test the null hypothesis that the differences are not significant. Our calculations are based on that assumption. If the differences do appear to be significant, we stop there. Not until the next and last chapter shall we attempt to analyze how differences are related.

Gorilla Heights

An anthropologist checked the heights of gorillas in two samples of 30, each sample from widely separated parts of West Africa. (He used a tranquilizer gun.) One sample showed an average height of 153.2 centimeters with a variance of 37.2 centimeters. The other showed an average of 162.6 centimeters with a variance of 40.9 centimeters. Does ANOVA suggest at a 0.05 level of significance that these two gorilla samples came from the same gorilla population?

$$H_0\text{: differences not significant}$$
$$H_1\text{: differences significant}$$

1. $\text{n.d.f.} = c - 1 = 2 - 1 = 1$

   $\text{d.d.f.} = c(r - 1) = 2(30 - 1) = 2(29) = 58$

   $\alpha = 0.05$

   In order to get F_e, **interpolation** is necessary. This means that, as no d.f. between 40 and 60 is given, you have to guess. First you find from the table that the F value for 40 is 4.08, and for 60 is 4.00. Say that 58 is $\frac{18}{20}$ of the way from 40 to 60. Then $\frac{18}{20}(4.08 - 4.00)$ produces 0.072: going 0.072 of the way from 4.08 to 4.00 gets you to 4.01.

$$F_e = 4.01$$

2. $s_1^2 = 37.2 \text{ cm} \qquad s_2^2 = 40.9 \text{ cm}$

3. $$\hat{\sigma}^2_{(\text{within})} = \frac{s_1^2 + s_2^2}{c} = \frac{37.2 + 40.9}{2}$$
   $$= \frac{78.1}{2} = 39.0 \text{ cm}$$

4. $$\hat{\mu}_{\bar{x}} = \frac{\bar{x}_1 + \bar{x}_2}{c} = \frac{153.2 + 162.6}{2}$$
   $$= \frac{315.8}{2} = 157.9$$

$$5.\ \hat{\sigma}^2_{\text{between}} = r\left[\frac{\sum(\bar{x} - \hat{\mu}_{\bar{x}})^2}{c - 1}\right]$$

$$= 30\left[\frac{(153.2 - 157.9)^2 + (162.6 - 157.9)^2}{2 - 1}\right]$$

$$= 30\left[\frac{(-4.7)^2 + 4.7^2}{1}\right]$$

$$= 30(22.09 + 22.09)$$

$$= 30(44.18) = 1325$$

$$6.\ F_a = \frac{\hat{\sigma}^2_{(\text{between})}}{\hat{\sigma}^2_{(\text{within})}}$$

$$= \frac{1325}{39}$$

$$= 34$$

Since F_e is 4.0 and F_a is 34, there is probably a significant difference between gorilla heights. Whether it is environment or heredity, we do not attempt to say. Simply

> Reject H_0: the differences in heights of gorillas from these two areas are significant.

15–6 SUMMARY

The analysis of variance furnishes us with a statistical technique for testing whether means are significantly different. You may use it to test one sample against a population hypothesis. You may use it to test any number of samples against a population hypothesis. And here you are not restricted to a sample size of 30 or more as you were with z and a normal distribution.

The technique requires looking up the maximum expected F for a given level of significance. Then the actual problem F is calculated as the ratio of the variance estimated from between samples over the variance estimated from within samples. Only if the actual F exceeds the expected F do we reject the null hypothesis, the null hypothesis being that the differences in means are due to chance and are not significant.

In short, if the variance between the different samples is large enough compared with the variance within the separate samples, it may indicate that the samples are significantly different. They may be different because of a relationship between them and their sources—different cobble banks, different gorilla species, different car makes, different machine operators.

Our final and climactic question is going to be: If we can show a significant relationship between paired variables, is it possible to measure it in such a way that, given any value of one member of a pair, we can predict the value of its mate? This will involve correlation and regression equations.

But first, try the exercises for this chapter. Are you making good use of Table VIII in the appendix? It will help you a great deal with ANOVA, too!

• IV •

STATISTICAL CONCEPTS FOR MEASURING RELATIONSHIPS

• 16 •

Regression and Correlation

But Galton was the pioneer in evolving the conception of a correlated system of variates, the representation "by a single numerical quantity of the degree of relationship, or of partial causality, between the different variables of our ever-changing universe."

JAMES R. NEWMAN
The World of Mathematics (1956)

OBJECTIVES

To see if there is a significant relationship
between paired variables,
and if there is,
to express this relationship as a linear equation.

SYMBOLS AND FORMULAS FIRST USED IN THIS CHAPTER:

$y = bx + a$	linear equation
$r = \dfrac{n\Sigma xy - \Sigma x \Sigma y}{\sqrt{n\Sigma x^2 - (\Sigma x)^2}\sqrt{n\Sigma y^2 - (\Sigma y)^2}}$	coefficient of correlation
$a = \dfrac{\Sigma y \Sigma x^2 - \Sigma x \Sigma xy}{n\Sigma x^2 - (\Sigma x)^2}$, $b = \dfrac{n\Sigma xy - \Sigma x \Sigma y}{n\Sigma x^2 - (\Sigma x)^2}$	constants in regression equation

16–1 EXERCISES

1. Do you expect a positive, negative, or zero correlation in the following?

   (a) World population and the year
   (b) Quantity of meat sold and the price
   (c) Expenditures for advertising and sales volume
   (d) Student hair color and grades
   (e) Heights of fathers and sons
   (f) Expense and duration of tourist trip
   (g) Football players' numbers and weight
   (h) Hat size and IQ
   (i) Refrigerator temperature and number at dial setting
   (j) Gasoline in tank and miles traveled

2. Explain what it means to have a correlation coefficient of $+1$, of 0, of -1. Give examples.

3. Realistic information where we are interested in correlation usually produces laborious computations, as you will see. Let's take a relatively easy one first.

TIME SPENT ON QUIZ

Minutes	Grade
6	3
8	3
10	4

The critical value of r will be fairly high because n is so small. Can you conclude that the correlation is significant at $\alpha = 0.05$? If it is, calculate an equation for the regression line.

4.

IQs OF FRATERNAL TWINS

	Male	Female
First pair	78	82
Second pair	106	112

Consider these two pairs of opposite-sexed fraternal twins (fraternal twins are from two separately fertilized eggs rather than, in the case of identical twins, from the splitting of a single fertilized egg). Could you predict the female IQs from the male IQs? ($\Sigma x = 184$, $\Sigma y = 194$, $\Sigma x^2 = 17{,}320$, $\Sigma y^2 = 19{,}268$, $\Sigma xy = 18{,}268$.) Calculate the coefficient of correlation. Don't two points determine a straight line? So what would you expect the correlation coefficient to be? Find the linear equation for a line through these two points. [Notice that when you substitute in the formulas for a and b, you have already found $n\Sigma x^2 - (\Sigma x)^2$ in determining the denominator for r; this now appears as the denominator for both a and b. Also, $n\Sigma xy - \Sigma x\Sigma y$, the numerator of r, is the numerator for b.]

5. Let's try it with a third pair.

	Male	**Female**
First pair	78	82
Second pair	106	112
Third pair	119	116

To save you arithmetic effort:

$$\Sigma x = 303$$
$$\Sigma y = 310$$
$$\Sigma x^2 = 31{,}481 = 31{,}480$$
$$\Sigma y^2 = 32{,}724 = 32{,}720$$
$$\Sigma xy = 32{,}072 = 32{,}070$$
$$(\Sigma x)^2 = 91{,}809 = 91{,}810$$
$$(\Sigma y)^2 = 96{,}100$$
$$\Sigma x \Sigma y = 93{,}930$$

Is your r significant at $\alpha = 0.05$? How about at $\alpha = 0.01$?

6. The heights of nine graduate students were checked against the heights of their fathers. The sample is small and the measurements not very precise, but let's go on to find the coefficient of correlation. It should appear to be significant at $\alpha = 0.05$, so calculate the regression equation. Use your regression equation to predict the father's height if the son's is 6′1″.

Students' Heights in Inches above 5 Feet	Corresponding Heights of Fathers
7	8
8	8
8	9
9	10
9	11
10	9
10	10
11	12
12	11

7. A sociologist claims that in a certain eastern seaboard city probably 1 out of 4 women married over two years will say in private that they are unhappily married. A skillfully contrived random sample of 200 women married over two years shows 40 as saying they are unhappily married. What could this indicate about the sociologist's claim at a 2% level of significance?

8. In shoe manufacturing, *lasting* is the process of forming the shoe over a solid foot form called a *last*. Is there a significant difference in the effectiveness of these four operators of automatic toe-lasting machines? Toes are put in cases of 36 pairs. Our results are in the number of cases. Production counts were taken on 5 consecutive days.

Operator A	Operator B	Operator C	Operator D
$n = 5$	$n = 5$	$n = 5$	$n = 5$
$\bar{x}_A = 29$	$\bar{x}_B = 30$	$\bar{x}_C = 30$	$\bar{x}_D = 31$
$s_A^2 = 1.9$	$s_B^2 = 1.6$	$s_C^2 = 2.0$	$s_D^2 = 0.9$

Use $\alpha = 0.05$.

9. An anthropologist claims that geographical location has an effect on the hair color of native children. At $\alpha = 0.01$, does the following information support his claim?

	Red Hair	Light Hair	Dark Hair
Maine	5	23	23
Louisiana	8	15	53
Colorado	38	38	50

10. A lunch-counter owner wants a 95% confidence interval estimate of the average customer expenditures for noon lunch, using $n = 100$, $\bar{x} =$ \$1.80, and $s =$ \$0.28.

11. For six winters in Rotterdam, Holland, a comparison of the number of days with a southerly (*Sz*) circulation and the corresponding mean precipitation is shown below. Make some specific statement about the significance of your calculated coefficient of correlation.

Winter	Number of Days with *Sz*	Mean Precipitation (mm)
1962–1963	13	304
1963–1964	3	269
1964–1965	2	472
1965–1966	0	492
1966–1967	2	498
1967–1968	0	468

12. The following bushels of grapes were used at a certain winery:

	Concord	Catawba
1960	3000	1000
1961	2900	1000
1962	3300	1200
1963	3000	1000
1964	3200	1100
1965	2700	1000

Is there a significant correlation between the use of Concord and Catawba grapes at $\alpha = 0.05$? Assume that the given figures have been rounded to two significant digits.

13. Do you suppose a student's grade in any subject would show a correlation with his average total weekly hours of study in all subjects? Here is a small sample.

Grade (F, D, C, B, A)	0	1	2	3	4
Average total weekly hours of study	2	5	8	14	21

If your correlation coefficient indicates a significant relationship, as it should at $\alpha = 0.05$, work out a regression equation. Try this equation on your own average weekly study time. What have you got to say about the soundness of this procedure for predicting?

14. A paper mill on the Nashua River claimed an average pollution index of 75 from its industrial waste. A 30-day test revealed an average daily pollution index of 80.1 with $s = 6.2$. What do these statistics suggest at a 1% level of significance?

15. The solubility of salts in water usually shows a curvilinear regression equation. Lead nitrate, $Pb(NO_3)_2$, however, seems to be linear. Find a correlation coefficient and regression equation suggested by the following information:

Temperature, (°C)	Grams of $Pb(NO_3)_2$ in 100 g of H_2O
0	37
10	47
20	57
30	67
40	77
50	86
60	95

16. An educator wants to estimate the IQs of children in the fourth grade of Michigan's public schools. He takes a random sample of 2000 and finds a sample mean of 105 with a sample standard deviation of 12. Make a 95% confidence interval estimate of the parametric mean.

16–2 NUMBER PAIRS AND THEIR RELATIONSHIPS

In this last chapter we are going to examine *related pairs* (the time it takes a body to fall and the distance it falls, the price of steel and the quantity sold, IQs of married males and their incomes, carrot lengths and their maximum diameters, Dow Jones averages of industrial stocks and money supply). Each pair has a measurement in one classification and a corresponding measurement in another classification (the pairs are usually presented as the separate rows of two columns). We are going to check the consistency of the relationship of these pairs, finding an index of this consistency called the **coefficient of correlation**. If the sample coefficient of correlation appears to be significant, we shall then attempt to work out an equation that will enable us, given one member of a pair, to predict the other member.

We have previously compared the proportions of different characteristics in samples from different sources (Chapter 14, Chi-Square) to see whether the characteristics from different sources were in significantly different proportions. We have just compared the means of different samples (Chapter 15, Analysis of Variance) to see if these sample means were significantly different. We are now, in effect, going to determine whether the differences between the members of pairs suggest a relationship that is significant.

We emphasize that, for the first time, we are not just going to say that there is or is not a significant relationship; we are going to attempt to measure that relationship (if it appears to be significant) in such a way that, given any value at all for one member of a pair, then we can predict the value of the corresponding member.

Actually we shall still be approaching our problem from a hypothesis-testing point of view, the null hypothesis being that there is not a significant relationship, the alternative hypothesis being, of course, that there is a significant relationship.

Again, our concern shall be with paired measurements. We cite several more pairs to impress upon you the wide possible application of correlation: education and salary, money in the bank and interest, family size and weekly expenditures, property owned and taxes paid, hours of study and grades, miles and gallons, electrical voltage and distances from the source, temperature of earth and depth of boring, molecular pressure and mass. The question always is: Do these paired variables show a relationship that is significantly more than we could expect from chance alone? If so, we try to express this relationship algebraically so that, given one member of a pair, we can predict the value of its partner.

Some theoretical relationships can be exactly expressed. The area of a circle is shown to be dependent upon the radius of the circle in exactly the same way every time. $A = \pi r^2$ (π is a constant). Interest on your loan (simple interest with the principal and rate constant) is predictably dependent upon the variable time, $i = prt$. These relationships are theoretically or arbitrarily fixed. Their models are algebraic equations.

But we knew the equations before we tried to find related pairs. When we meet up with pairs for which we have no equations, we have to decide first whether they are related and, if they are, then try to work out possible equations for this relationship.

A usual approach to newly presented pairs is to graph them. We proceed here to plot on graph paper five sets of pairs. In four of the cases, we show the actual algebraic model in the form $y = bx + a$ (a is the intersection of the line with the vertical axis; b is the slope of the line). These lines either have a perfect correlation (+1 or −1) when all points fall on the line, or they have a correlation somewhere in between +1 and −1 when the points do not all fall on a straight line.

Again, when all of the points fall on a straight line, we show the coefficient of correlation as 1 (+1 if the line rises from left to right, as in the case of tree age and tree height, −1 if the line falls from left to right, as with miles traveled and gas in the tank). A positive correlation indicates a direct relationship (when one variable increases, the other increases). A negative correlation indicates an inverse relationship (when one variable increases, the other decreases). If the points do not fall on a straight line, the index of how close they come to it is the coefficient of correlation and takes on any value from +1 to −1. A coefficient of ± (plus or minus) 0.9 would be good. A coefficient of ± 0.1 would be poor. A coefficient of 0 would indicate no correlation whatever. Actually showing no correlation whatever is probably a rarity. Even hip measurements and IQs would probably not show a correlation coefficient of exactly 0; that is why it is so very important to establish a level of significance.

All linear equations basically have the algebraic form $y = bx + a$. We now proceed to plot five sets of paired variables showing their equations and correlation coefficients, when possible.

(1)

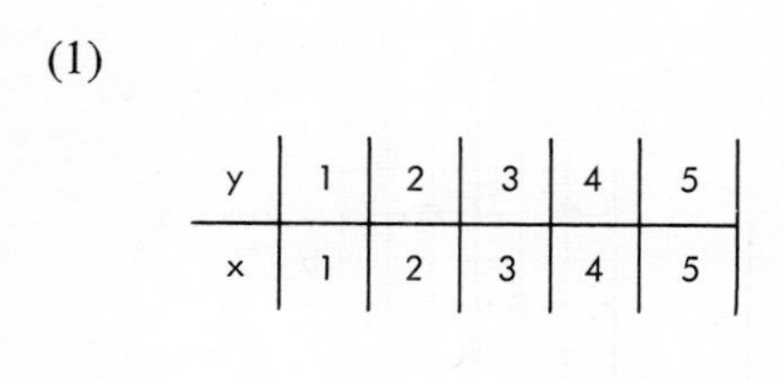

y	1	2	3	4	5
x	1	2	3	4	5

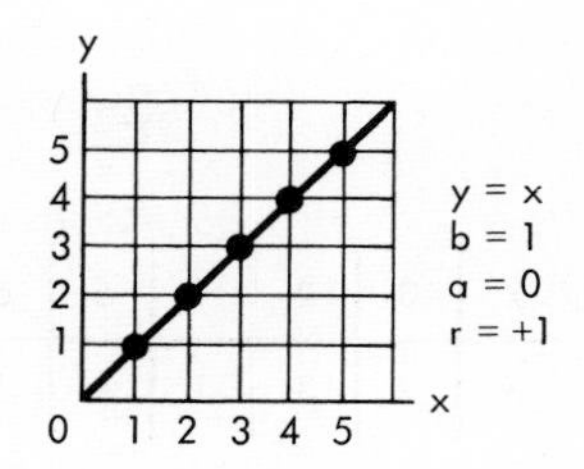

(2)

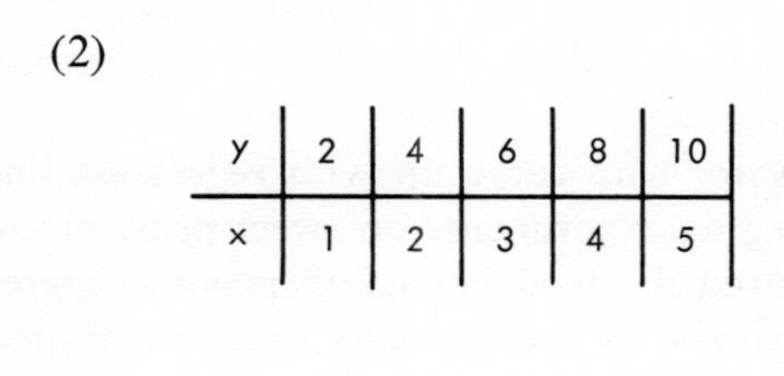

y	2	4	6	8	10
x	1	2	3	4	5

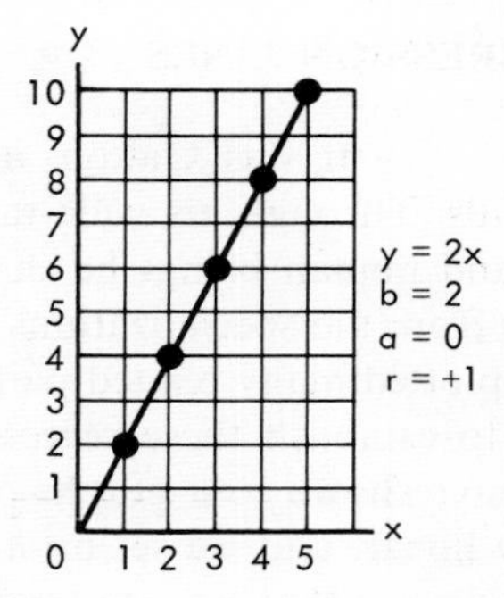

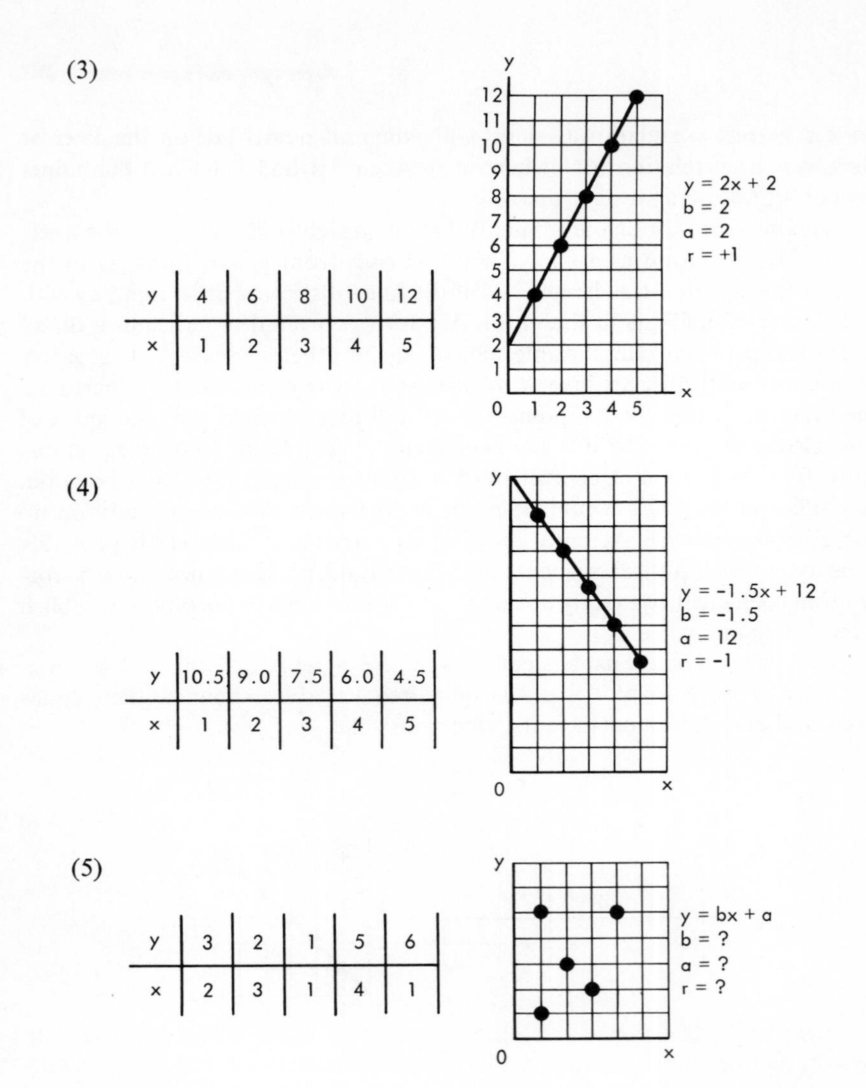

(3)

y	4	6	8	10	12
x	1	2	3	4	5

(4)

y	10.5	9.0	7.5	6.0	4.5
x	1	2	3	4	5

(5)

y	3	2	1	5	6
x	2	3	1	4	1

16–3 REGRESSION LINES

It was Galton, again, who came up with **regression lines** in the 1870s. Through his wide-ranging experiments on sweet peas, moths, hounds, and human beings he showed the tendency of offspring to regress (go back) from the specializations shown by their parents to a sort of mean type. He plotted many related pairs, such as the heights of fathers and sons, in order to establish these regression equations.

We have shown four graphs of related pairs where the graphs are quite obviously linear, and we set up an algebraic equation for each line. In the fifth case we see that no one straight line can be drawn through all of the points.

Of course sometimes we can get a line to pass through *some* of the points. Here we show the vertical distances of all our points from our trial line.

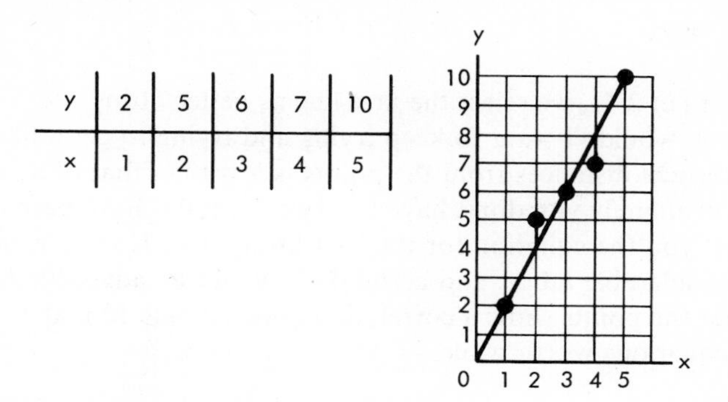

y	2	5	6	7	10
x	1	2	3	4	5

You could try by eye, or by a ruler, or by a piece of string to get a better fit. Is your line a better fit?

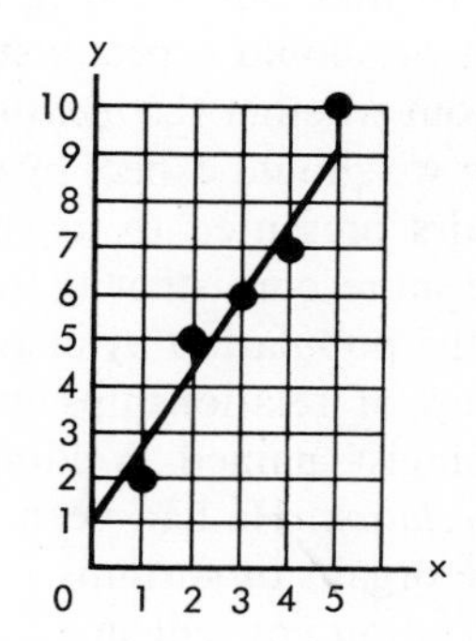

Keep moving the line around. How do you know when you have the best-fitting line? You can answer this question more confidently by the method of least squares than you can by eye. Measure the vertical distances of all points from the assumed line (if the point is on the line, the distance from it is zero, of course) and then square all the distances and add them. The line which has the least sum of these squares will fit the best. This establishes the regression line.

In our first try (see graph), we square the vertical distances from the line and add them.

$$0^2 + 1^2 + 0^2 + 1^2 + 0^2 = 2$$

In our second try:

$$\left(\frac{2}{3}\right)^2 + \left(\frac{2}{3}\right)^2 + 0^2 + \left(\frac{1}{2}\right)^2 + 1^2 = \frac{4}{9} + \frac{4}{9} + 0 + \frac{1}{4} + 1$$

$$= \frac{16 + 16 + 0 + 9 + 36}{36}$$

$$= \frac{77}{36} = 2.14$$

As 2 is less than 2.14, we select the first line as better fitting.

Now you wouldn't want to keep trying and trying to get a line the sum of whose vertical distances from the points is less than that of any previous line. And fortunately you don't have to! Two formulas have been developed that will get you the equation for the best fitting line. However, using them involves considerable labor, and certainly it would be advisable to find out first whether the points show a correlation good enough to make calculating the linear equations worth while.

16–4 CORRELATION COEFFICIENTS

In chi-square we tried to find out whether *counts* from a population showed more variation than we should expect just by chance. In the analysis of variance we tried to find out whether the *measurements* from a population showed more variation than we should expect by chance. Now we are trying to find out whether the pairs presented to us from a population of pairs show relationships that are more consistent than we should expect from a pairing of the members of the population by chance.

This sort of relationship of relationships was called *correlation* by a late nineteenth-century biologist named Weldon. He also designed what is called a *coefficient of correlation.* He had been studying the related variation in measurements of the organs of shrimps.

Let's consider the correlation concept in respect to GPAs (grade point averages). Do student GPAs in high school show a significant correlation with their GPAs as freshmen in college?

Here's how a random sample appears in a table:

Student	High-School GPA	College GPA
1	3.10	2.60
2	3.60	3.00
3	2.70	2.50
4	2.20	1.80
5	3.00	2.80
6	2.70	3.20
7	3.50	2.80
8	3.50	2.00
9	2.00	1.80
10	3.50	3.30

The pairs are plotted on graph paper as shown in the following figure.

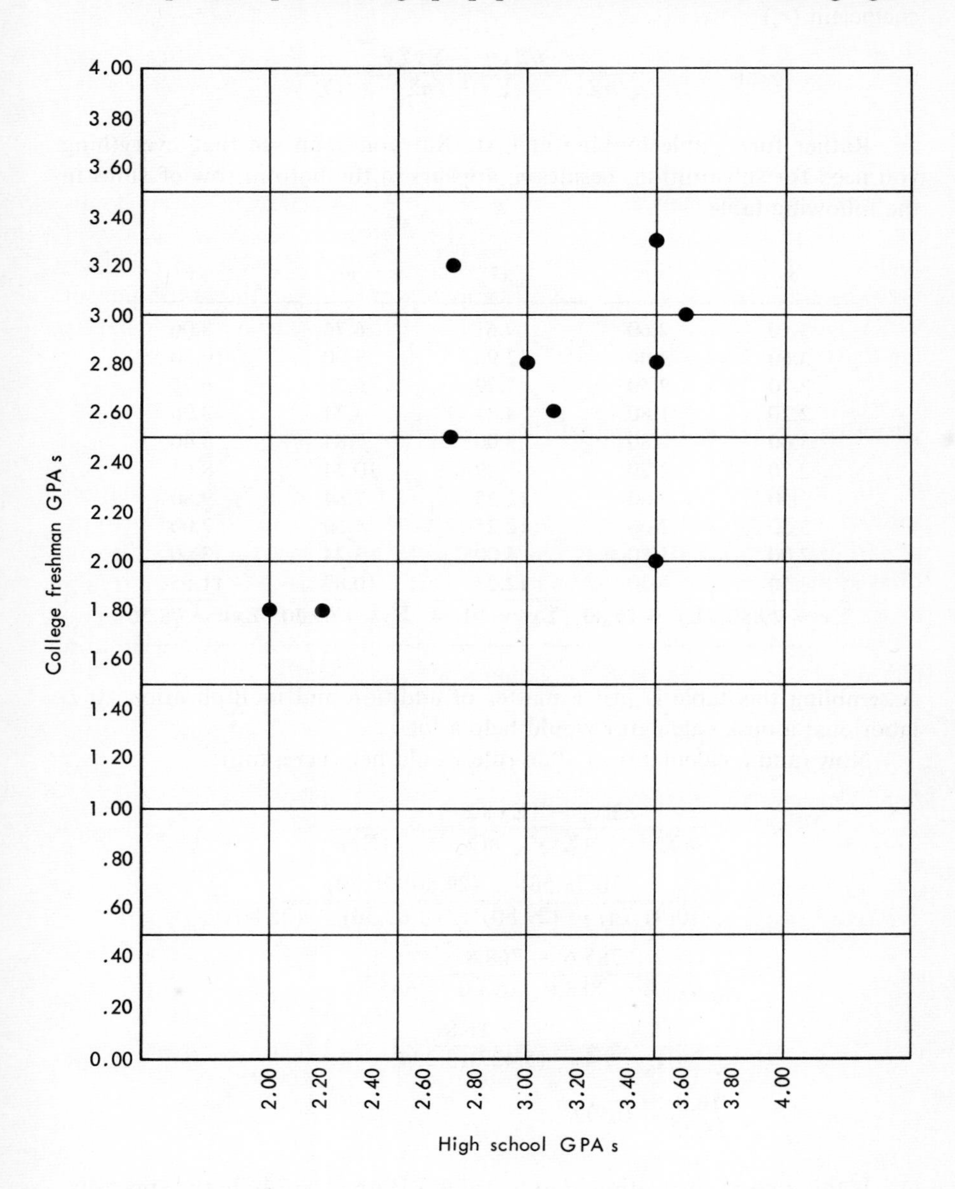

You could try for the best fit with an elastic band. Our regression-line formula, however, will do a better job for you. As I said, before you do all this work on a regression equation, it would be to your advantage to find out whether the coefficient of correlation indicates that you have any right to expect that the correlation is significant. So, for your actual correlation coefficient (r_a)

$$r_a = \frac{n\Sigma xy - \Sigma x \Sigma y}{\sqrt{n\Sigma x^2 - (\Sigma x)^2}\sqrt{n\Sigma y^2 - (\Sigma y)^2}}$$

Rather formidable looking at first! But you soon see that everything you need for substitution, besides n, appears in the bottom row of sums in the following table.

x	y	x^2	y^2	xy
3.10	2.60	9.61	6.76	8.06
3.60	3.00	12.96	9.00	10.80
2.70	2.50	7.29	6.25	6.75
2.20	1.80	4.84	3.24	3.96
3.00	2.80	9.00	7.84	8.40
2.70	3.20	7.29	10.24	8.64
3.50	2.80	12.25	7.84	9.80
3.50	2.00	12.25	4.00	7.00
2.00	1.80	4.00	3.24	3.60
3.50	3.30	12.25	10.89	11.55
$\Sigma x = 29.80$	$\Sigma y = 25.80$	$\Sigma x^2 = 91.74$	$\Sigma y^2 = 69.30$	$\Sigma xy = 78.56$

Assembling this table is just a matter of addition and multiplication. It *is* laborious; a desk calculator would help a lot.

Now (and a calculator or slide rule would help here, too):

$$r = \frac{n\Sigma xy - (\Sigma x)(\Sigma y)}{\sqrt{n(\Sigma x^2) - (\Sigma x)^2}\sqrt{n(\Sigma y^2) - (\Sigma y)^2}}$$

$$= \frac{10(78.56) - (29.80)(25.80)}{\sqrt{10(91.74) - (29.80)^2}\sqrt{10(69.30) - (25.80)^2}}$$

$$= \frac{785.6 - 768.8}{\sqrt{917.4 - 888.0}\sqrt{693.0 - 665.6}}$$

$$= \frac{16.8}{\sqrt{29.4}\sqrt{27.4}} = \frac{16.8}{(5.422)(5.234)}$$

$$= \frac{16.8}{28.38} = 0.592$$

Is this a good correlation? Go to Table VII on page 242 in the appendix. Here are listed critical values for r, the coefficient of correlation. If your calculated coefficient of correlation exceeds the appropriate critical value, you may go about calculating a regression equation, happy with the conviction that the correlation is probably significant. How do you get the appropriate critical r? Knowing the degrees of freedom to be $n - 2$ and $\alpha = 0.05$ or $\alpha = 0.01$, you certainly ought to be able by yourself to find out how the table works.

Comparing your calculated correlation coefficient ($r = 0.592$) with the critical correlation coefficient at d.f. $= 8$ and $\alpha = 0.01$ ($r_\alpha = 0.765$), you find that you do not have a significant correlation. So why calculate a regression equation?

There are two concepts to emphasize here. First, the level of significance concept. When anyone applies to correlation a qualitative adjective like *good, important, meaningful*, you've got to ask: good, important, meaningful at what level of significance? Statistical inference demands that you indicate numerically the likelihood of a Type I error.

Second, the chance element in paired variables. If you use first two pairs, you will always get perfect correlation because the pairs supply us with just two points (one for each pair) and the two points determine a straight line. Even three pairs are likely to exhibit a false correlation just by chance. As the sample size of pairs increases, however, it becomes less and less likely that the pairs will show a correlation that does not really exist. If you compare 100 pairs and if there isn't a fairly high correlation in the whole population of pairs, you are very unlikely to get one in the sample. This danger of a false correlation from a small sample is also something to avoid.

Here's a case where the coefficient of correlation will show a significant relationship. A cardiac stimulant measured in grains (1 grain is about 65 milligrams) and administered to a patient at intervals of one week, produces the shown increase in number of heartthrobs per minute.

TABLE OF PAIRS

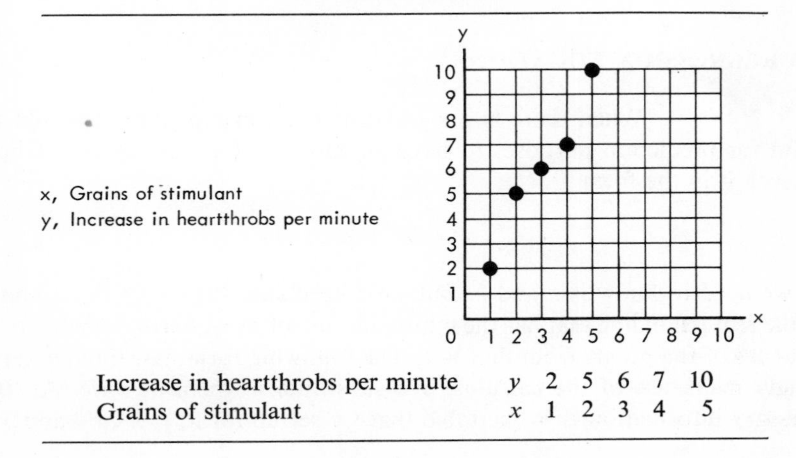

Increase in heartthrobs per minute	y	2	5	6	7	10
Grains of stimulant	x	1	2	3	4	5

TABLE FOR CALCULATIONS

x	y	x^2	y^2	xy
1	2	1	4	2
2	5	4	25	10
3	6	9	36	18
4	7	16	49	28
5	10	25	100	50
$\Sigma x = 15$	$\Sigma y = 30$	$\Sigma x^2 = 55$	$\Sigma y^2 = 214$	$\Sigma xy = 108$

Again, all the substitutions come from our table of calculations:

$$r = \frac{n\Sigma xy - \Sigma x \Sigma y}{\sqrt{n\Sigma x^2 - (\Sigma x)^2}\sqrt{n\Sigma y^2 - (\Sigma y)^2}}$$

$$= \frac{5(108) - (15)(30)}{\sqrt{5(55) - (15)^2}\sqrt{5(214) - (30)^2}}$$

$$= \frac{540 - 450}{\sqrt{275 - 225}\sqrt{1070 - 900}}$$

$$= \frac{90}{\sqrt{50}\sqrt{170}} = \frac{90}{(7.07)(13.0)} = \frac{90.0}{91.9} = 0.98$$

Using $\alpha = 0.01$ and d.f. $= n - 2 = 5 - 2 = 3$ from Table VII, we get $r_\alpha = 0.959$. This result is less than our actual 0.98. The coefficient of correlation for our heartthrob problem exceeds the coefficient we could expect if only chance were at work. So we say that there *is* a significant relationship between the amount of stimulant administered and the increase in heartthrob count.

16–5 REGRESSION EQUATIONS

When there is a significant relationship, then it would be useful for prediction purposes to have an equation for this regression line. We seek it in the form

$$y = bx + a$$

All we need to know is a and b. But we'd need calculus to get the a and b for the regression line that has the minimum sum of the squares of the vertical distances of the points from that line. The following formulas, though, have already made use of the calculus. We substitute in them at once. All the necessary information is in the table that we set up for x, y, x^2, y^2, and xy.

$$a = \frac{\Sigma y \Sigma x^2 - \Sigma x \Sigma xy}{n\Sigma x^2 - (\Sigma x)^2}$$

$$= \frac{(30)(55) - (15)(108)}{5(55) - (15)^2}$$

$$= \frac{1650 - 1620}{275 - 225} = \frac{30}{50} = 0.6$$

$$b = \frac{n\Sigma xy - \Sigma x \Sigma y}{n\Sigma x^2 - (\Sigma x)^2}$$

$$= \frac{5(108) - (15)(30)}{5(55) - (15)^2}$$

$$= \frac{540 - 450}{275 - 225} = \frac{90}{50} = 1.8$$

From these calculations emerges a valuable regression equation

$$y = 1.8x + 0.6$$

Substitute any value for x and you will get a corresponding value for y. However, predicting from within our data is much safer than going outside the data. For example, 1.5 grains of the stimulant might produce a heartthrob increase of

$$y = 1.8x + 0.6 = 1.8(1.5) + 0.6$$
$$= 2.7 + 0.6 = 3.3 \text{ (increase in heartthrobs per minute)}$$

But this increase in rate of throbs after administering 15 grains of the stimulant (way outside our empirical limits) might be very much less than our equation predicts—because the patient is probably dead! We have, nevertheless, found a fairly good means of predicting *within* the limits of our information. Eureka!
spurious. First, there is the matter of going beyond the given pairs of information. For instance, you could probably show that over the last 20 years the average number of people in motor cars has been decreasing. A regression equation might suggest the feasibility of going *beyond* the figures, the feasability of *extrapolating* But coming up with the prediction that by 1980 every other car would have nobody in it would certainly be unrealistic.

There is also the danger of assuming the correlation to indicate a cause-and-effect relationship. It might be possible to show that increased candy eating shows a positive correlation with sexual desire. If you measure candy consumed in grams per week and rate the corresponding sexual urge in millivolts, it might be possible to predict anywhere within the measured range. (Extrapolating here could be pretty dull or pretty exciting!) But the relationship may not be a causal one; it may simply be that people who have a lot of time for candy have a lot of time for daydreaming. Correlation and regression equations have great potential, but to use them you'd better know what you are doing.

This presentation of the calculation of a regression equation and the determination of two correlation coefficients should, with a little solid application on your part, enable you to do the correlation problems that appear in our final exercise. (You should be able to handle all of the review problems, too.)

16–6 CONCLUSION

Given drug dosages and corresponding heartthrob increases, we calculated their coefficient of correlation. Then, since the coefficient of correlation suggested a significant relationship, we found a regression equation. Consequently, given the value of one variable not appearing in our original data, we actually predicted the value of the other one to go with it. Thus, we not only measured the significance of a relationship; we also stated algebraically and numerically what that relationship might be. This is the climax to our story. And if you have hung on this far, you know that the climax means a great deal more than the few words on this page that state it.

But, still, if you look around, you may see even greater complexities than any we have touched upon. Actual functions are not always, or even relatively often, expressible as straight line functions. In *curvilinear analysis* there are exponential curves (like population growth), hyperbolic curves (like supply and demand), and many others. But both simple linear and curvilinear analysis involve two variables in isolation from any others. The equation for the relationship between the radius of a circle and its area, completely free from other influences, is $A = \pi r^2$, like $y = \pi x^2$. This relationship has a correlation coefficient of $+1$. It is a beautiful theoretical model—always correct up there in mathematical heaven, never exactly so down here on empirical earth.

Actually, down here everything affects everything else in varying degrees. We live in a multivariant world. A reliable analysis of most situations requires this multivariate point of view, an extension of our work on two-variable correlation and regression. The manufacturer of anything—chain saws to chocolate bars—is usually not satisfied with how one variable affects his product and consequently his income. He wants to know how a combination of variables affects them. This may call for a specialist in *multivariate analysis* (and the pay is good).

Multivariate analysis, curvilinear functions, decision theory, experimental design, game theory, nonparametric testing, rank correlation—maybe you'd better take another statistics course. Or you could pursue some independent study on your own.

Anyway, write the American Statistical Association, 806 15th Street, NW, Washington, D.C., 20005, for its pamphlet, *Careers in Statistics*.

Everywhere about us we find variability and uncertainty and the shimmering suggestion of relationships. Among such circumstances man has lived for centuries, complicating his political structures, increasing his sociological problems, polluting his natural environment, using up his energy sources—basing his decisions on insight, intuition, and want. But this is not good enough. He needs a more scientific, systematic, sophisticated approach to decision making. And the accomplishment of this end will be greatly facilitated by better sampling techniques and refinements in the basic statistical concepts of estimation, testing hypotheses, and measuring relationships.

As for causes—who knows? "The cause is hidden, but the result is known." [Ovid, A.D. 18]

Appendix

TABLE I RANDOM NUMBERS

02946	96520	81881	56247	17623	47441	27821	91845
85697	62000	87957	07258	45054	58410	92081	97624
26734	68426	52067	23123	73700	58730	06111	64486
47829	32353	95941	72169	58374	03905	06865	95353
76603	99339	40571	41186	04981	17531	97372	39558
47526	26522	11045	83565	45639	02485	43905	01823
70100	85732	19741	92951	98832	38188	24080	24519
86819	50200	50889	06493	66638	03619	90906	95370
41614	30074	23403	03656	77580	87772	86877	57085
17930	26194	53836	53692	67125	98175	00912	11246
24649	31845	25736	75231	83808	98997	71829	99430
79899	34061	54308	59358	56462	58166	97302	86828
76801	49594	81002	30397	52728	15101	72070	33706
62567	08480	61873	63162	44873	35302	04511	38088
49723	15275	09399	11211	67352	41526	23497	75440
42658	70183	89417	57676	35370	14915	16569	54945
65080	35569	79392	14937	06081	74957	87787	68849
02906	38119	72407	71427	58478	99297	43519	62410
75153	86376	63852	60557	21211	77299	74967	99038
14192	49525	78844	13664	98964	64425	33536	15079
32059	11548	86264	74406	81496	23996	56872	71401
81716	80301	96704	57214	71361	41989	92589	69788
43315	50483	02950	09611	36341	20326	37489	34626
27510	10769	09921	46721	34183	22856	18724	60422
81782	04769	36716	82519	98272	13969	12429	03093
19975	48346	91029	78902	75689	70722	88553	83300
98356	76855	18769	52843	64204	95212	31320	03783
29708	17814	31556	68610	16574	42305	56300	84227
88014	27583	78167	25057	93552	74363	30951	41367
94491	19238	17396	10592	48907	79840	34607	62668
56957	05072	53948	07850	42569	82391	20435	79306
50915	31924	80621	17495	81618	15125	48087	01250
49631	93771	80200	84622	31413	33756	15218	81976
99683	58162	45516	39761	77600	15175	67415	88801
86017	20264	94618	85979	42009	78616	45210	73186
77339	64605	82583	85011	02955	84348	46436	77911
61714	57933	37342	26000	93611	93346	71212	24405
15232	48027	15832	62924	11509	95853	02747	61889
41447	34275	10779	83515	63899	30932	90572	98971
23244	43524	16382	36340	73581	76780	03842	64009
53460	83542	25224	70378	49604	14809	12317	78062
53442	16897	61578	05032	81825	76822	87170	77235
55548	19096	04130	23104	60534	44842	16954	99466
18185	69329	02340	63111	41788	74409	76177	55519
02372	45690	38595	23121	73818	74454	02371	94693
51715	35492	61371	87132	81585	55439	98095	55578
24717	16786	42786	86985	21858	39489	39251	70450
78022	32604	87259	93702	99438	68184	62119	20229
35995	08275	62405	43313	03249	74135	43003	63132
29192	86922	31908	42703	59638	31226	89860	45191

TABLE I RANDOM NUMBERS (*cont.*)

01654	50375	23941	44848	79154	30193	15271	93296
73750	68343	40727	81203	91727	06463	12248	57567
64163	22132	22896	14305	45642	71580	21558	66457
88445	85544	23627	79176	32502	34568	78777	35179
94180	71108	19121	11958	33308	75930	24865	17426
11433	12220	36719	35435	77727	78493	94580	70091
61838	68801	49856	21739	95701	11180	73936	17628
59908	68103	36855	19127	56637	06383	83182	76927
41018	69556	06402	03436	89803	95604	55735	66978
85095	01581	92299	06166	71078	31823	64316	95567
59705	78103	66740	41743	69177	59277	47629	63874
75094	55208	77905	20705	32408	16630	01242	63119
78425	68672	79455	94334	23292	90422	75540	15843
89088	86918	20787	05691	97309	96673	16389	51437
83345	95889	39333	86027	24680	51909	97230	58136
62896	00342	66647	57096	84913	67895	18804	17691
96498	38270	80532	54307	07885	38892	50990	96766
30974	47335	04918	42974	19294	72581	77377	04652
57901	06163	99162	53285	27341	02507	41858	08436
88494	80633	47785	53996	57058	04222	54488	12019
34883	00045	89682	86664	92195	42593	56488	35402
24373	46438	28935	63903	14722	10715	58795	42800
16828	79262	23678	05509	23733	95318	77730	87614
33723	27646	92335	87136	88062	21506	01750	71326
01542	75066	73921	97188	31250	41996	31680	41783
00100	12787	74100	95536	42359	01761	28842	71562
82697	03389	19303	21646	22532	81701	03425	28914
28137	17549	22698	72955	59849	02370	02784	13711
02248	21570	33796	83789	72981	96423	68791	91684
56175	82515	23348	42207	87644	57353	90349	16448
80020	21622	67659	07878	17586	65524	20162	04712
20271	23094	48372	77621	32889	19595	66500	28064
38734	98044	02658	90698	72563	15076	23780	52815
48183	24263	49297	32923	94406	63865	44336	27224
48163	34158	03177	51696	57795	31725	14403	29856
45658	15024	66664	18730	40671	92727	68626	81631
71128	15524	55666	14763	13729	51708	54104	81331
19041	42899	49464	93965	14960	88896	72784	82054
32672	67506	93040	94527	31556	80163	80203	90928
15823	48310	04391	15521	79255	69253	60254	01653
82810	18981	62581	31642	42693	78972	60322	90462
74772	80840	05816	29023	67410	12916	87933	78840
52931	38199	85632	23761	99084	48028	07184	41635
95395	87644	09722	99251	97129	70847	91864	08549
76695	33451	57139	90612	11918	90871	60965	23555
83560	50374	04410	57272	36705	51302	93147	29479
28355	62002	85994	35807	84810	14186	51153	78998
84684	54861	41330	66808	65231	14168	45193	27156
21135	92001	43896	55887	35319	03793	60344	95970
24236	01536	43897	41294	45551	46877	58631	82654

TABLE II BINOMIAL PROBABILITIES

		π										
n	x	**0.050**	**0.100**	**0.200**	**0.300**	**0.400**	**0.500**	**0.600**	**0.700**	**0.800**	**0.900**	**0.950**
2	0	0.902	0.810	0.640	0.490	0.360	0.250	0.160	0.090	0.040	0.010	0.002
	1	0.095	0.180	0.320	0.420	0.480	0.500	0.480	0.420	0.320	0.180	0.095
	2	0.002	0.010	0.040	0.090	0.160	0.250	0.360	0.490	0.640	0.810	0.902
3	0	0.857	0.729	0.512	0.343	0.216	0.125	0.064	0.027	0.008	0.001	
	1	0.135	0.243	0.384	0.441	0.432	0.375	0.288	0.189	0.096	0.027	0.007
	2	0.007	0.027	0.096	0.189	0.288	0.375	0.432	0.441	0.384	0.243	0.135
	3		0.001	0.008	0.027	0.064	0.125	0.216	0.343	0.512	0.729	0.857
4	0	0.815	0.656	0.410	0.240	0.130	0.062	0.026	0.008	0.002		
	1	0.171	0.292	0.410	0.412	0.346	0.250	0.154	0.076	0.026	0.004	
	2	0.014	0.049	0.154	0.265	0.346	0.375	0.346	0.265	0.154	0.049	0.014
	3		0.004	0.026	0.076	0.154	0.250	0.346	0.412	0.410	0.292	0.171
	4			0.002	0.008	0.026	0.062	0.130	0.240	0.410	0.656	0.815
5	0	0.774	0.590	0.328	0.168	0.078	0.031	0.010	0.002			
	1	0.204	0.328	0.410	0.360	0.259	0.156	0.077	0.028	0.006		
	2	0.021	0.073	0.205	0.309	0.346	0.312	0.230	0.132	0.051	0.008	0.001
	3	0.001	0.008	0.051	0.132	0.230	0.312	0.346	0.309	0.205	0.073	0.021
	4			0.006	0.028	0.077	0.156	0.259	0.360	0.410	0.328	0.204
	5				0.002	0.010	0.031	0.078	0.168	0.328	0.590	0.774
6	0	0.735	0.531	0.262	0.118	0.047	0.016	0.004	0.001			
	1	0.232	0.354	0.393	0.303	0.187	0.094	0.037	0.010	0.002		
	2	0.031	0.098	0.246	0.324	0.311	0.234	0.138	0.060	0.015	0.001	
	3	0.002	0.015	0.082	0.185	0.276	0.312	0.276	0.185	0.082	0.015	0.002
	4		0.001	0.015	0.060	0.138	0.234	0.311	0.324	0.246	0.098	0.031
	5			0.002	0.010	0.037	0.094	0.187	0.303	0.393	0.354	0.232
	6				0.001	0.004	0.016	0.047	0.118	0.262	0.531	0.735
7	0	0.698	0.478	0.210	0.082	0.028	0.008	0.002				
	1	0.257	0.372	0.367	0.247	0.131	0.055	0.017	0.004			
	2	0.041	0.124	0.275	0.318	0.261	0.164	0.077	0.025	0.004		
	3	0.004	0.023	0.115	0.227	0.290	0.273	0.194	0.097	0.029	0.003	
	4		0.003	0.029	0.097	0.194	0.273	0.290	0.227	0.115	0.023	0.004
	5			0.004	0.025	0.077	0.164	0.261	0.318	0.275	0.124	0.041
	6				0.004	0.017	0.055	0.131	0.247	0.367	0.372	0.257
	7					0.002	0.008	0.028	0.082	0.210	0.478	0.698
8	0	0.663	0.430	0.168	0.058	0.017	0.004	0.001				
	1	0.279	0.383	0.336	0.198	0.090	0.031	0.008	0.001			
	2	0.051	0.149	0.294	0.296	0.209	0.109	0.041	0.010	0.001		
	3	0.005	0.033	0.147	0.254	0.279	0.219	0.124	0.047	0.009		
	4		0.005	0.046	0.136	0.232	0.273	0.232	0.136	0.046	0.005	
	5			0.009	0.047	0.124	0.219	0.279	0.254	0.147	0.033	0.005
	6			0.001	0.010	0.041	0.109	0.209	0.296	0.294	0.149	0.051
	7				0.001	0.008	0.031	0.090	0.198	0.336	0.383	0.279
	8					0.001	0.004	0.017	0.058	0.168	0.430	0.663

TABLE II BINOMIAL PROBABILITIES (*cont.*)

		π										
n	x	**0.050**	**0.100**	**0.200**	**0.300**	**0.400**	**0.500**	**0.600**	**0.700**	**0.800**	**0.900**	**0.950**
9	0	0.630	0.387	0.134	0.040	0.010	0.002					
	1	0.299	0.387	0.302	0.156	0.060	0.018	0.004				
	2	0.063	0.172	0.302	0.267	0.161	0.070	0.021	0.004			
	3	0.008	0.045	0.176	0.267	0.251	0.164	0.074	0.021	0.003		
	4	0.001	0.007	0.066	0.172	0.251	0.246	0.167	0.074	0.017	0.001	
	5		0.001	0.017	0.074	0.167	0.246	0.251	0.172	0.066	0.007	0.001
	6			0.003	0.021	0.074	0.164	0.251	0.267	0.176	0.045	0.008
	7				0.004	0.021	0.070	0.161	0.267	0.302	0.172	0.063
	8					0.004	0.018	0.060	0.156	0.302	0.387	0.299
	9						0.002	0.010	0.040	0.134	0.387	0.630
10	0	0.599	0.349	0.107	0.028	0.006	0.001					
	1	0.315	0.387	0.268	0.121	0.040	0.010	0.002				
	2	0.075	0.194	0.302	0.233	0.121	0.044	0.011	0.001			
	3	0.010	0.057	0.201	0.267	0.215	0.117	0.042	0.009	0.001		
	4	0.001	0.011	0.088	0.200	0.251	0.205	0.111	0.037	0.006		
	5		0.001	0.026	0.103	0.201	0.246	0.201	0.103	0.026	0.001	
	6			0.006	0.037	0.111	0.205	0.251	0.200	0.088	0.011	0.001
	7			0.001	0.009	0.042	0.117	0.215	0.267	0.201	0.057	0.010
	8				0.001	0.011	0.044	0.121	0.233	0.302	0.194	0.075
	9					0.002	0.010	0.040	0.121	0.268	0.387	0.315
	10						0.001	0.006	0.028	0.107	0.349	0.599
11	0	0.569	0.314	0.086	0.020	0.004						
	1	0.329	0.384	0.236	0.093	0.027	0.005	0.001				
	2	0.087	0.213	0.295	0.200	0.089	0.027	0.005	0.001			
	3	0.014	0.071	0.221	0.257	0.177	0.081	0.023	0.004			
	4	0.001	0.016	0.111	0.220	0.236	0.161	0.070	0.017	0.002		
	5		0.002	0.039	0.132	0.221	0.226	0.147	0.057	0.010		
	6			0.010	0.057	0.147	0.226	0.221	0.132	0.039	0.002	
	7			0.002	0.017	0.070	0.161	0.236	0.220	0.111	0.016	0.001
	8				0.004	0.023	0.081	0.177	0.257	0.221	0.071	0.014
	9				0.001	0.005	0.027	0.089	0.200	0.295	0.213	0.087
	10					0.001	0.005	0.027	0.093	0.236	0.384	0.329
	11							0.004	0.020	0.086	0.314	0.569
12	0	0.540	0.282	0.069	0.014	0.002						
	1	0.341	0.377	0.206	0.071	0.017	0.003					
	2	0.099	0.230	0.283	0.168	0.064	0.016	0.002				
	3	0.017	0.085	0.236	0.240	0.142	0.054	0.012	0.001			
	4	0.002	0.021	0.133	0.231	0.213	0.121	0.042	0.008	0.001		
	5		0.004	0.053	0.158	0.227	0.193	0.101	0.029	0.003		
	6			0.016	0.079	0.177	0.226	0.177	0.079	0.016		
	7			0.003	0.029	0.101	0.193	0.227	0.158	0.053	0.004	
	8			0.001	0.008	0.042	0.121	0.213	0.231	0.133	0.021	0.002
	9				0.001	0.012	0.054	0.142	0.240	0.236	0.085	0.017
	10					0.002	0.016	0.064	0.168	0.283	0.230	0.099
	11						0.003	0.017	0.071	0.206	0.377	0.341
	12							0.002	0.014	0.069	0.282	0.540

TABLE II BINOMIAL PROBABILITIES (*cont.*)

		π										
n	*x*	**0.050**	**0.100**	**0.200**	**0.300**	**0.400**	**0.500**	**0.600**	**0.700**	**0.800**	**0.900**	**0.950**
13	0	0.513	0.254	0.055	0.010	0.001						
	1	0.351	0.367	0.179	0.054	0.011	0.002					
	2	0.111	0.245	0.268	0.139	0.045	0.010	0.001				
	3	0.021	0.100	0.246	0.218	0.111	0.035	0.006	0.001			
	4	0.003	0.028	0.154	0.234	0.184	0.087	0.024	0.003			
	5		0.006	0.069	0.180	0.221	0.157	0.066	0.014	0.001		
	6		0.001	0.023	0.103	0.197	0.209	0.131	0.044	0.006		
	7			0.006	0.044	0.131	0.209	0.197	0.103	0.023	0.001	
	8			0.001	0.014	0.066	0.157	0.221	0.180	0.069	0.006	
	9				0.003	0.024	0.087	0.184	0.234	0.154	0.028	0.003
	10				0.001	0.006	0.035	0.111	0.218	0.246	0.100	0.021
	11					0.001	0.010	0.045	0.139	0.268	0.245	0.111
	12						0.002	0.011	0.054	0.179	0.367	0.351
	13							0.001	0.010	0.055	0.254	0.513
14	0	0.488	0.229	0.044	0.007	0.001						
	1	0.359	0.356	0.154	0.041	0.007	0.001					
	2	0.123	0.257	0.250	0.113	0.032	0.006	0.001				
	3	0.026	0.114	0.250	0.194	0.085	0.022	0.003				
	4	0.004	0.035	0.172	0.229	0.155	0.061	0.014	0.001			
	5		0.008	0.086	0.196	0.207	0.122	0.041	0.007			
	6		0.001	0.032	0.126	0.207	0.183	0.092	0.023	0.002		
	7			0.009	0.062	0.157	0.209	0.157	0.062	0.009		
	8			0.002	0.023	0.092	0.183	0.207	0.126	0.032	0.001	
	9				0.007	0.041	0.122	0.207	0.196	0.086	0.008	
	10				0.001	0.014	0.061	0.155	0.229	0.172	0.035	0.004
	11					0.003	0.022	0.085	0.194	0.250	0.114	0.026
	12					0.001	0.006	0.032	0.113	0.250	0.257	0.123
	13						0.001	0.007	0.041	0.154	0.356	0.359
	14							0.001	0.007	0.044	0.229	0.488
15	0	0.463	0.206	0.035	0.005							
	1	0.366	0.343	0.132	0.031	0.005						
	2	0.135	0.267	0.231	0.092	0.022	0.003					
	3	0.031	0.129	0.250	0.170	0.063	0.014	0.002				
	4	0.005	0.043	0.188	0.219	0.127	0.042	0.007	0.001			
	5	0.001	0.010	0.103	0.206	0.186	0.092	0.024	0.003			
	6		0.002	0.043	0.147	0.207	0.153	0.061	0.012	0.001		
	7			0.014	0.081	0.177	0.196	0.118	0.035	0.003		
	8			0.003	0.035	0.118	0.196	0.177	0.081	0.014		
	9			0.001	0.012	0.061	0.153	0.207	0.147	0.043	0.002	
	10				0.003	0.024	0.092	0.186	0.206	0.103	0.010	0.001
	11				0.001	0.007	0.042	0.127	0.219	0.188	0.043	0.005
	12					0.002	0.014	0.063	0.170	0.250	0.129	0.031
	13						0.003	0.022	0.092	0.231	0.267	0.135
	14							0.005	0.031	0.132	0.343	0.366
	15								0.005	0.035	0.206	0.463

TABLE III DIGIT SEQUENCES FOR SQUARE ROOTS

100	1000	3162	150	1225	3873	200	1414	4472
101	1005	3178	151	1229	3886	201	1418	4483
102	1010	3194	152	1233	3899	202	1421	4494
103	1015	3209	153	1237	3912	203	1425	4506
104	1020	3225	154	1241	3924	204	1428	4517
105	1025	3240	155	1245	3937	205	1432	4528
106	1030	3256	156	1249	3950	206	1435	4539
107	1034	3271	157	1253	3962	207	1439	4550
108	1039	3286	158	1257	3975	208	1442	4561
109	1044	3302	159	1261	3987	209	1446	4572
110	1049	3317	160	1265	4000	210	1449	4583
111	1054	3332	161	1269	4012	211	1453	4593
112	1058	3347	162	1273	4025	212	1456	4604
113	1063	3362	163	1277	4037	213	1459	4615
114	1068	3376	164	1281	4050	214	1463	4626
115	1072	3391	165	1285	4062	215	1466	4637
116	1077	3406	166	1288	4074	216	1470	4648
117	1082	3421	167	1292	4087	217	1473	4658
118	1086	3435	168	1296	4099	218	1476	4669
119	1091	3450	169	1300	4111	219	1480	4680
120	1095	3464	170	1304	4123	220	1483	4690
121	1100	3479	171	1308	4135	221	1487	4701
122	1105	3493	172	1311	4147	222	1490	4712
123	1109	3507	173	1315	4159	223	1493	4722
124	1114	3521	174	1319	4171	224	1497	4733
125	1118	3536	175	1323	4183	225	1500	4743
126	1122	3550	176	1327	4195	226	1503	4754
127	1127	3564	177	1330	4207	227	1507	4764
128	1131	3578	178	1334	4219	228	1510	4775
129	1136	3592	179	1338	4231	229	1513	4785
130	1140	3606	180	1342	4243	230	1517	4796
131	1145	3619	181	1345	4254	231	1520	4806
132	1149	3633	182	1349	4266	232	1523	4817
133	1153	3647	183	1353	4278	233	1526	4827
134	1158	3661	184	1356	4290	234	1530	4837
135	1162	3674	185	1360	4301	235	1533	4848
136	1166	3688	186	1364	4313	236	1536	4858
137	1170	3701	187	1367	4324	237	1539	4868
138	1175	3715	188	1371	4336	238	1543	4879
139	1179	3728	189	1375	4347	239	1546	4889
140	1183	3742	190	1378	4359	240	1549	4899
141	1187	3755	191	1382	4370	241	1552	4909
142	1192	3768	192	1386	4382	242	1556	4919
143	1196	3782	193	1389	4393	243	1559	4930
144	1200	3795	194	1393	4405	244	1562	4940
145	1204	3808	195	1396	4416	245	1565	4950
146	1208	3821	196	1400	4427	246	1568	4960
147	1212	3834	197	1404	4438	247	1572	4970
148	1217	3847	198	1407	4450	248	1575	4980
149	1221	3860	199	1411	4461	249	1578	4990

TABLE III DIGIT SEQUENCES FOR SQUARE ROOTS (*cont.*)

250	1581	5000	300	1732	5477	350	1871	5916
251	1584	5010	301	1735	5486	351	1873	5925
252	1587	5020	302	1738	5495	352	1876	5933
253	1591	5030	303	1741	5505	353	1879	5941
254	1594	5040	304	1744	5514	354	1881	5950
255	1597	5050	305	1746	5523	355	1884	5958
256	1600	5060	306	1749	5532	356	1887	5967
257	1603	5070	307	1752	5541	357	1889	5975
258	1606	5079	308	1755	5550	358	1892	5983
259	1609	5089	309	1758	5559	359	1895	5992
260	1612	5099	310	1761	5568	360	1897	6000
261	1616	5109	311	1764	5577	361	1900	6008
262	1619	5119	312	1766	5586	362	1903	6017
263	1622	5128	313	1769	5595	363	1905	6025
264	1625	5138	314	1772	5604	364	1908	6033
265	1628	5148	315	1775	5612	365	1910	6042
266	1631	5158	316	1778	5621	366	1913	6050
267	1634	5167	317	1780	5630	367	1916	6058
268	1637	5177	318	1783	5639	368	1918	6066
269	1640	5187	319	1786	5648	369	1921	6075
270	1643	5196	320	1789	5657	370	1924	6083
271	1646	5206	321	1792	5666	371	1926	6091
272	1649	5215	322	1794	5675	372	1929	6099
273	1652	5225	323	1797	5683	373	1931	6107
274	1655	5234	324	1800	5692	374	1934	6116
275	1658	5244	325	1803	5701	375	1936	6124
276	1661	5254	326	1806	5710	376	1939	6132
277	1664	5263	327	1808	5718	377	1942	6140
278	1667	5273	328	1811	5727	378	1944	6148
279	1670	5282	329	1814	5736	379	1947	6156
280	1673	5292	330	1817	5745	380	1949	6164
281	1676	5301	331	1819	5753	381	1952	6173
282	1679	5310	332	1822	5762	382	1954	6181
283	1682	5320	333	1825	5771	383	1957	6189
284	1685	5329	334	1828	5779	384	1960	6197
285	1688	5339	335	1830	5788	385	1962	6205
286	1691	5348	336	1833	5797	386	1965	6213
287	1694	5357	337	1836	5805	387	1967	6221
288	1697	5367	338	1838	5814	388	1970	6229
289	1700	5376	339	1841	5822	389	1972	6237
290	1703	5385	340	1844	5831	390	1975	6245
291	1706	5394	341	1847	5840	391	1977	6253
292	1709	5404	342	1849	5848	392	1980	6261
293	1712	5413	343	1852	5857	393	1982	6269
294	1715	5422	344	1855	5865	394	1985	6277
295	1718	5431	345	1857	5874	395	1987	6285
296	1720	5441	346	1860	5882	396	1990	6293
297	1723	5450	347	1863	5891	397	1992	6301
298	1726	5459	348	1865	5899	398	1995	6309
299	1729	5468	349	1868	5908	399	1997	6317

TABLE III DIGIT SEQUENCES FOR SQUARE ROOTS (*cont.*)

400	2000	6325	450	2121	6708	500	2236	7071
401	2002	6332	451	2124	6716	501	2238	7078
402	2005	6340	452	2126	6723	502	2241	7085
403	2007	6348	453	2128	6731	503	2243	7092
404	2010	6356	454	2131	6738	504	2245	7099
405	2012	6364	455	2133	6745	505	2247	7106
406	2015	6372	456	2135	6753	506	2249	7113
407	2017	6380	457	2138	6760	507	2252	7120
408	2020	6387	458	2140	6768	508	2254	7127
409	2022	6395	459	2142	6775	509	2256	7134
410	2025	6403	460	2145	6782	510	2258	7141
411	2027	6411	461	2147	6790	511	2261	7148
412	2030	6419	462	2149	6797	512	2263	7155
413	2032	6427	463	2152	6804	513	2265	7162
414	2035	6434	464	2154	6812	514	2267	7169
415	2037	6442	465	2156	6819	515	2269	7176
416	2040	6450	466	2159	6826	516	2272	7183
417	2042	6458	467	2161	6834	517	2274	7190
418	2045	6465	468	2163	6841	518	2276	7197
419	2047	6473	469	2166	6848	519	2278	7204
420	2049	6481	470	2168	6856	520	2280	7211
421	2052	6488	471	2170	6863	521	2283	7218
422	2054	6496	472	2173	6870	522	2285	7225
423	2057	6504	473	2175	6877	523	2287	7232
424	2059	6512	474	2177	6885	524	2289	7239
425	2062	6519	475	2179	6892	525	2291	7246
426	2064	6527	476	2182	6899	526	2293	7253
427	2066	6535	477	2184	6907	527	2296	7259
428	2069	6542	478	2186	6914	528	2298	7266
429	2071	6550	479	2189	6921	529	2300	7273
430	2074	6557	480	2191	6928	530	2302	7280
431	2076	6565	481	2193	6935	531	2304	7287
432	2078	6573	482	2195	6943	532	2307	7294
433	2081	6580	483	2198	6950	533	2309	7301
434	2083	6588	484	2200	6957	534	2311	7308
435	2086	6595	485	2202	6964	535	2313	7314
436	2088	6603	486	2205	6971	536	2315	7321
437	2090	6611	487	2207	6979	537	2317	7328
438	2093	6618	488	2209	6986	538	2319	7335
439	2095	6626	489	2211	6993	539	2322	7342
440	2098	6633	490	2214	7000	540	2324	7348
441	2100	6641	491	2216	7007	541	2326	7355
442	2102	6648	492	2218	7014	542	2328	7362
443	2105	6656	493	2220	7021	543	2330	7369
444	2107	6663	494	2223	7029	544	2332	7376
445	2110	6671	495	2225	7036	545	2335	7382
446	2112	6678	496	2227	7043	546	2337	7389
447	2114	6686	497	2229	7050	547	2339	7396
448	2117	6693	498	2232	7057	548	2341	7403
449	2119	6701	499	2234	7064	549	2343	7409

TABLE III DIGIT SEQUENCES FOR SQUARE ROOTS (*cont.*)

550	2345	7416
551	2347	7423
552	2349	7430
553	2352	7436
554	2354	7443
555	2356	7450
556	2358	7457
557	2360	7463
558	2362	7470
559	2364	7477
560	2366	7483
561	2369	7490
562	2371	7497
563	2373	7503
564	2375	7510
565	2377	7517
566	2379	7523
567	2381	7530
568	2383	7537
569	2385	7543
570	2387	7550
571	2390	7556
572	2392	7563
573	2394	7570
574	2396	7576
575	2398	7583
576	2400	7589
577	2402	7596
578	2404	7603
579	2406	7609
580	2408	7616
581	2410	7622
582	2412	7629
583	2415	7635
584	2417	7642
585	2419	7649
586	2421	7655
587	2423	7662
588	2425	7668
589	2427	7675
590	2429	7681
591	2431	7688
592	2433	7694
593	2435	7701
594	2437	7707
595	2439	7714
596	2441	7720
597	2443	7727
598	2445	7733
599	2447	7740
600	2449	7746
601	2452	7752
602	2454	7759
603	2456	7765
604	2458	7772
605	2460	7778
606	2462	7785
607	2464	7791
608	2466	7797
609	2468	7804
610	2470	7810
611	2472	7817
612	2474	7823
613	2476	7829
614	2478	7836
615	2480	7842
616	2482	7849
617	2484	7855
618	2486	7861
619	2488	7868
620	2490	7874
621	2492	7880
622	2494	7887
623	2496	7893
624	2498	7899
625	2500	7906
626	2502	7912
627	2504	7918
628	2506	7925
629	2508	7931
630	2510	7937
631	2512	7944
632	2514	7950
633	2516	7956
634	2518	7962
635	2520	7969
636	2522	7975
637	2524	7981
638	2526	7987
639	2528	7994
640	2530	8000
641	2532	8006
642	2534	8012
643	2536	8019
644	2538	8025
645	2540	8031
646	2542	8037
647	2544	8044
648	2546	8050
649	2548	8056
650	2550	8062
651	2551	8068
652	2553	8075
653	2555	8081
654	2557	8087
655	2559	8093
656	2561	8099
657	2563	8106
658	2565	8112
659	2567	8118
660	2569	8124
661	2571	8130
662	2573	8136
663	2575	8142
664	2577	8149
665	2579	8155
666	2581	8161
667	2583	8167
668	2585	8173
669	2587	8179
670	2588	8185
671	2590	8191
672	2592	8198
673	2594	8204
674	2596	8210
675	2598	8216
676	2600	8222
677	2602	8228
678	2604	8234
679	2606	8240
680	2608	8246
681	2610	8252
682	2612	8258
683	2613	8264
684	2615	8270
685	2617	8276
686	2619	8283
687	2621	8289
688	2623	8295
689	2625	8301
690	2627	8307
691	2629	8313
692	2631	8319
693	2632	8325
694	2634	8331
695	2636	8337
696	2638	8343
697	2640	8349
698	2642	8355
699	2644	8361

TABLE III DIGIT SEQUENCES FOR SQUARE ROOTS (*cont.*)

700	2646	8367	750	2739	8660	800	2828	8944
701	2648	8373	751	2740	8666	801	2830	8950
702	2650	8379	752	2742	8672	802	2832	8955
703	2651	8385	753	2744	8678	803	2834	8961
704	2653	8390	754	2746	8683	804	2835	8967
705	2655	8396	755	2748	8689	805	2837	8972
706	2657	8402	756	2750	8695	806	2839	8978
707	2659	8408	757	2751	8701	807	2841	8983
708	2661	8414	758	2753	8706	808	2843	8989
709	2663	8420	759	2755	8712	809	2844	8994
710	2665	8426	760	2757	8718	810	2846	9000
711	2666	8432	761	2759	8724	811	2848	9006
712	2668	8438	762	2760	8729	812	2850	9011
713	2670	8444	763	2762	8735	813	2851	9017
714	2672	8450	764	2764	8741	814	2853	9022
715	2674	8456	765	2766	8746	815	2855	9028
716	2676	8462	766	2768	8752	816	2857	9033
717	2678	8468	767	2769	8758	817	2858	9039
718	2680	8473	768	2771	8764	818	2860	9044
719	2681	8479	769	2773	8769	819	2862	9050
720	2683	8485	770	2775	8775	820	2864	9055
721	2685	8491	771	2777	8781	821	2865	9061
722	2687	8497	772	2778	8786	822	2867	9066
723	2689	8503	773	2780	8792	823	2869	9072
724	2691	8509	774	2782	8798	824	2871	9077
725	2693	8515	775	2784	8803	825	2872	9083
726	2694	8521	776	2786	8809	826	2874	9088
727	2696	8526	777	2787	8815	827	2876	9094
728	2698	8532	778	2789	8820	828	2877	9099
729	2700	8538	779	2791	8826	829	2879	9105
730	2702	8544	780	2793	8832	830	2881	9110
731	2704	8550	781	2795	8837	831	2883	9116
732	2706	8556	782	2796	8843	832	2884	9121
733	2707	8562	783	2798	8849	833	2886	9127
734	2709	8567	784	2800	8854	834	2888	9132
735	2711	8573	785	2802	8860	835	2890	9138
736	2713	8579	786	2804	8866	836	2891	9143
737	2715	8585	787	2805	8871	837	2893	9149
738	2717	8591	788	2807	8877	838	2895	9154
739	2718	8597	789	2809	8883	839	2897	9160
740	2720	8602	790	2811	8888	840	2898	9165
741	2722	8608	791	2812	8894	841	2900	9171
742	2724	8614	792	2814	8899	842	2902	9176
743	2726	8620	793	2816	8905	843	2903	9182
744	2728	8626	794	2818	8911	844	2905	9187
745	2729	8631	795	2820	8916	845	2907	9192
746	2731	8637	796	2821	8922	846	2909	9198
747	2733	8643	797	2823	8927	847	2910	9203
748	2735	8649	798	2825	8933	848	2912	9209
749	2737	8654	799	2827	8939	849	2914	9214

TABLE III DIGIT SEQUENCES FOR SQUARE ROOTS (*cont.*)

850	2915	9220	900	3000	9487	950	3082	9747
851	2917	9225	901	3002	9492	951	3084	9752
852	2919	9230	902	3003	9497	952	3085	9757
853	2921	9236	903	3005	9503	953	3087	9762
854	2922	9241	904	3007	9508	954	3089	9767
855	2924	9247	905	3008	9513	955	3090	9772
856	2926	9252	906	3010	9518	956	3092	9778
857	2927	9257	907	3012	9524	957	3094	9783
858	2929	9263	908	3013	9529	958	3095	9788
859	2931	9268	909	3015	9534	959	3097	9793
860	2933	9274	910	3017	9539	960	3098	9798
861	2934	9279	911	3018	9545	961	3100	9803
862	2936	9284	912	3020	9550	962	3102	9808
863	2938	9290	913	3022	9555	963	3103	9813
864	2939	9295	914	3023	9560	964	3105	9818
865	2941	9301	915	3025	9566	965	3106	9823
866	2943	9306	916	3027	9571	966	3108	9829
867	2944	9311	917	3028	9576	967	3110	9834
868	2946	9317	918	3030	9581	968	3111	9839
869	2948	9322	919	3031	9586	969	3113	9844
870	2950	9327	920	3033	9592	970	3114	9849
871	2951	9333	921	3035	9597	971	3116	9854
872	2953	9338	922	3036	9602	972	3118	9859
873	2955	9343	923	3038	9607	973	3119	9864
874	2956	9349	924	3040	9612	974	3121	9869
875	2958	9354	925	3041	9618	975	3122	9874
876	2960	9359	926	3043	9623	976	3124	9879
877	2961	9365	927	3045	9628	977	3126	9884
878	2963	9370	928	3046	9633	978	3127	9889
879	2965	9375	929	3048	9638	979	3129	9894
880	2966	9381	930	3050	9644	980	3130	9899
881	2968	9386	931	3051	9649	981	3132	9905
882	2970	9391	932	3053	9654	982	3134	9910
883	2972	9397	933	3055	9659	983	3135	9915
884	2973	9402	934	3056	9644	984	3137	9920
885	2975	9407	935	3058	9670	985	3138	9925
886	2977	9413	936	3059	9675	986	3140	9930
887	2978	9418	937	3061	9680	987	3142	9935
888	2980	9423	938	3063	9685	988	3143	9940
889	2982	9429	939	3064	9690	989	3145	9945
890	2983	9434	940	3066	9695	990	3146	9950
891	2985	9439	941	3068	9701	991	3148	9955
892	2987	9445	942	3069	9706	992	3150	9960
893	2988	9450	943	3071	9711	993	3151	9965
894	2990	9455	944	3072	9716	994	3153	9970
895	2992	9460	945	3074	9721	995	3154	9975
896	2993	9466	946	3076	9726	996	3156	9980
897	2995	9471	947	3077	9731	997	3158	9985
898	2997	9476	948	3079	9737	998	3159	9990
899	2998	9482	949	3081	9742	999	3161	9995

TABLE IV(a) NORMAL CURVE AREAS BETWEEN MEAN AND z

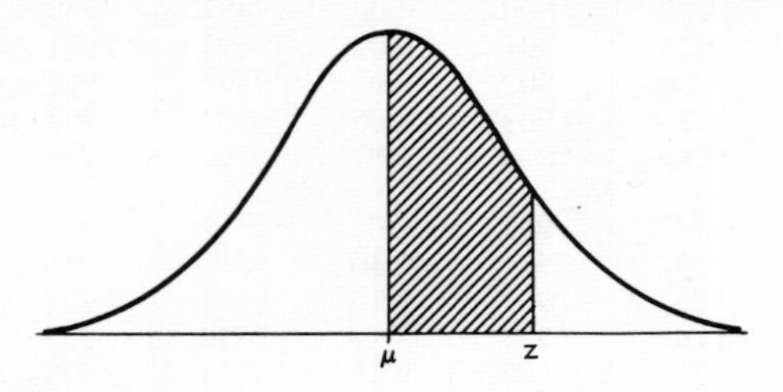

z	0.00	0.01	0.02	0.03	0.04	0.05	0.06	0.07	0.08	0.09
0.0	.0000	.0040	.0080	.0120	.0160	.0199	.0239	.0279	.0319	.0359
0.1	.0398	.0438	.0478	.0517	.0557	.0596	.0636	.0675	.0714	.0753
0.2	.0793	.0832	.0871	.0910	.0948	.0987	.1026	.1064	.1103	.1141
0.3	.1179	.1217	.1255	.1293	.1331	.1368	.1406	.1443	.1480	.1517
0.4	.1554	.1591	.1628	.1664	.1700	.1736	.1772	.1808	.1844	.1879
0.5	.1915	.1950	.1985	.2019	.2054	.2088	.2123	.2157	.2190	.2224
0.6	.2257	.2291	.2324	.2357	.2389	.2422	.2454	.2486	.2517	.2549
0.7	.2580	.2611	.2642	.2673	.2704	.2734	.2764	.2794	.2823	.2852
0.8	.2881	.2910	.2939	.2967	.2995	.3023	.3051	.3078	.3105	.3133
0.9	.3159	.3186	.3212	.3238	.3264	.3289	.3315	.3340	.3365	.3389
1.0	.3413	.3438	.3461	.3485	.3508	.3531	.3554	.3577	.3599	.3621
1.1	.3643	.3665	.3686	.3708	.3729	.3749	.3770	.3790	.3810	.3830
1.2	.3849	.3869	.3888	.3907	.3925	.3944	.3962	.3980	.3997	.4015
1.3	.4032	.4049	.4066	.4082	.4099	.4115	.4131	.4147	.4162	.4177
1.4	.4192	.4207	.4222	.4236	.4251	.4265	.4279	.4292	.4306	.4319
1.5	.4332	.4345	.4357	.4370	.4382	.4394	.4406	.4418	.4429	.4441
1.6	.4452	.4463	.4474	.4484	.4495	.4505	.4515	.4525	.4535	.4545
1.7	.4554	.4564	.4573	.4582	.4591	.4599	.4608	.4616	.4625	.4633
1.8	.4641	.4649	.4656	.4664	.4671	.4678	.4686	.4693	.4699	.4706
1.9	.4713	.4719	.4726	.4732	.4738	.4744	.4750	.4756	.4761	.4767
2.0	.4772	.4778	.4783	.4788	.4793	.4798	.4803	.4808	.4812	.4817
2.1	.4821	.4826	.4830	.4834	.4838	.4842	.4846	.4850	.4854	.4857
2.2	.4861	.4864	.4868	.4871	.4875	.4878	.4881	.4884	.4887	.4890
2.3	.4893	.4896	.4898	.4901	.4904	.4906	.4909	.4911	.4913	.4916
2.4	.4918	.4920	.4922	.4925	.4927	.4929	.4931	.4932	.4934	.4936
2.5	.4938	.4940	.4941	.4943	.4945	.4946	.4948	.4949	.4951	.4952
2.6	.4953	.4955	.4956	.4957	.4959	.4960	.4961	.4962	.4963	.4964
2.7	.4965	.4966	.4967	.4968	.4969	.4970	.4971	.4972	.4973	.4974
2.8	.4974	.4975	.4976	.4977	.4977	.4978	.4979	.4979	.4980	.4981
2.9	.4981	.4982	.4982	.4983	.4984	.4984	.4985	.4985	.4986	.4986
3.0	.4987	.4987	.4987	.4988	.4988	.4989	.4989	.4989	.4990	.4990

TABLE IV(b) NORMAL CURVE AREAS BEYOND z

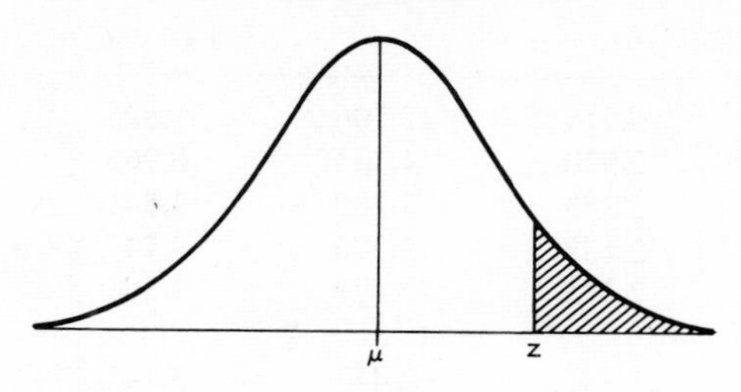

z	0.00	0.01	0.02	0.03	0.04	0.05	0.06	0.07	0.08	0.09
0.0	.5000	.4960	.4920	.4880	.4840	.4801	.4761	.4721	.4681	.4641
0.1	.4602	.4562	.4522	.4483	.4443	.4404	.4364	.4325	.4286	.4247
0.2	.4207	.4168	.4129	.4090	.4052	.4013	.3974	.3936	.3897	.3859
0.3	.3821	.3783	.3745	.3707	.3669	.3632	.3594	.3557	.3520	.3483
0.4	.3446	.3409	.3372	.3336	.3300	.3264	.3228	.3192	.3156	.3121
0.5	.3085	.3050	.3015	.2981	.2946	.2912	.2877	.2843	.2810	.2776
0.6	.2743	.2709	.2676	.2643	.2611	.2578	.2546	.2514	.2483	.2451
0.7	.2420	.2389	.2358	.2327	.2296	.2266	.2236	.2206	.2177	.2148
0.8	.2119	.2090	.2061	.2033	.2005	.1977	.1949	.1922	.1894	.1867
0.9	.1841	.1814	.1788	.1762	.1736	.1711	.1685	.1660	.1635	.1611
1.0	.1587	.1562	.1539	.1515	.1492	.1469	.1446	.1423	.1401	.1379
1.1	.1357	.1335	.1314	.1292	.1271	.1251	.1230	.1210	.1190	.1170
1.2	.1151	.1131	.1112	.1093	.1075	.1056	.1038	.1020	.1003	.0985
1.3	.0968	.0951	.0934	.0918	.0901	.0885	.0869	.0853	.0838	.0823
1.4	.0808	.0793	.0778	.0764	.0749	.0735	.0721	.0708	.0694	.0681
1.5	.0668	.0655	.0643	.0630	.0618	.0606	.0594	.0582	.0571	.0559
1.6	.0548	.0537	.0526	.0516	.0505	.0495	.0485	.0475	.0465	.0455
1.7	.0446	.0436	.0427	.0418	.0409	.0401	.0392	.0384	.0375	.0367
1.8	.0359	.0351	.0344	.0336	.0329	.0322	.0314	.0307	.0301	.0294
1.9	.0287	.0281	.0274	.0268	.0262	.0256	.0250	.0244	.0239	.0233
2.0	.0228	.0222	.0217	.0212	.0207	.0202	.0197	.0192	.0188	.0183
2.1	.0179	.0174	.0170	.0166	.0162	.0158	.0154	.0150	.0146	.0143
2.2	.0139	.0136	.0132	.0129	.0125	.0122	.0119	.0116	.0113	.0110
2.3	.0107	.0104	.0102	.0099	.0096	.0094	.0091	.0089	.0087	.0084
2.4	.0082	.0080	.0078	.0075	.0073	.0071	.0069	.0068	.0066	.0064
2.5	.0062	.0060	.0059	.0057	.0055	.0054	.0052	.0051	.0049	.0048
2.6	.0047	.0045	.0044	.0043	.0041	.0040	.0039	.0038	.0037	.0036
2.7	.0035	.0034	.0033	.0032	.0031	.0030	.0029	.0028	.0027	.0026
2.8	.0026	.0025	.0024	.0023	.0023	.0022	.0021	.0021	.0020	.0019
2.9	.0019	.0018	.0018	.0017	.0016	.0016	.0015	.0015	.0014	.0014
3.0	.0013	.0013	.0013	.0012	.0012	.0011	.0011	.0011	.0010	.0010

TABLE IV(c) THE t DISTRIBUTION*

d.f.	$t_{.100}$	$t_{.050}$	$t_{.025}$	$t_{.010}$	$t_{.005}$	d.f.
1	3.078	6.314	12.706	31.821	63.657	1
2	1.886	2.920	4.303	6.965	9.925	2
3	1.638	2.353	3.182	4.541	5.841	3
4	1.533	2.132	2.776	3.747	4.604	4
5	1.476	2.015	2.571	3.365	4.032	5
6	1.440	1.943	2.447	3.143	3.707	6
7	1.415	1.895	2.365	2.998	3.499	7
8	1.397	1.860	2.306	2.896	3.355	8
9	1.383	1.833	2.262	2.821	3.250	9
10	1.372	1.812	2.228	2.764	3.169	10
11	1.363	1.796	2.201	2.718	3.106	11
12	1.356	1.782	2.179	2.681	3.055	12
13	1.350	1.771	2.160	2.650	3.012	13
14	1.345	1.761	2.145	2.624	2.977	14
15	1.341	1.753	2.131	2.602	2.947	15
16	1.337	1.746	2.120	2.583	2.921	16
17	1.333	1.740	2.110	2.567	2.898	17
18	1.330	1.734	2.101	2.552	2.878	18
19	1.328	1.729	2.093	2.539	2.861	19
20	1.325	1.725	2.086	2.528	2.845	20
21	1.323	1.721	2.080	2.518	2.831	21
22	1.321	1.717	2.074	2.508	2.819	22
23	1.319	1.714	2.069	2.500	2.807	23
24	1.318	1.711	2.064	2.492	2.797	24
25	1.316	1.708	2.060	2.485	2.787	25
26	1.315	1.706	2.056	2.479	2.779	26
27	1.314	1.703	2.052	2.473	2.771	27
28	1.313	1.701	2.048	2.467	2.763	28
29	1.311	1.699	2.045	2.462	2.756	29
inf.	1.282	1.645	1.960	2.326	2.576	inf.

*This table is taken from Table IV of R.A. Fisher, *Statistical Methods for Research Workers*, published by Longman Group Ltd., London (previously published by Oliver & Boyd, Edinburgh 1963), and by permission of the authors and publishers.

TABLE V CHI-SQUARE DISTRIBUTION*

d.f.	$\chi^2_{.05}$	$\chi^2_{.01}$	d.f.†
1	3.841	6.635	1
2	5.991	9.210	2
3	7.815	11.345	3
4	9.488	13.277	4
5	11.070	15.086	5
6	12.592	16.812	6
7	14.067	18.475	7
8	15.507	20.090	8
9	16.919	21.666	9
10	18.307	23.209	10
11	19.675	24.725	11
12	21.026	26.217	12
13	22.362	27.688	13
14	23.685	29.141	14
15	24.996	30.578	15
16	26.296	32.000	16
17	27.587	33.409	17
18	28.869	34.805	18
19	30.144	36.191	19
20	31.410	37.566	20
21	32.671	38.932	21
22	33.924	40.289	22
23	35.172	41.638	23
24	36.415	42.980	24
25	37.652	44.314	25
26	38.885	45.642	26
27	40.113	46.963	27
28	41.337	48.278	28
29	42.557	49.588	29
30	43.773	50.892	30

*This table is based on Table 8 of *Biometrika Tables for Statisticians,* Vol. I (New York: Cambridge University Press, 1966) by permission of the *Biometrika* trustees.

†d.f. = $(r - 1)(c - 1)$

TABLE VI (a) F RATIOS, $\alpha = 0.05$*

d.f. for Denominator	Degrees of Freedom for Numerator 1	2	3	4	5	6	7	8	9	10	12	15	20	24	30	40	60	120	M
1	161	200	216	225	230	234	237	239	241	242	244	246	248	249	250	251	252	253	254
2	18.5	19.0	19.2	19.2	19.3	19.3	19.4	19.4	19.4	19.4	19.4	19.4	19.4	19.5	19.5	19.5	19.5	19.5	19.5
3	10.1	9.55	9.28	9.12	9.01	8.94	8.89	8.85	8.81	8.79	8.74	8.70	8.66	8.64	8.62	8.59	8.57	8.55	8.53
4	7.71	6.94	6.59	6.39	6.26	6.16	6.09	6.04	6.00	5.96	5.91	5.86	5.80	5.77	5.75	5.72	5.69	5.66	5.63
5	6.61	5.79	5.41	5.19	5.05	4.95	4.88	4.82	4.77	4.74	4.68	4.62	4.56	4.53	4.50	4.46	4.43	4.40	4.37
6	5.99	5.14	4.76	4.53	4.39	4.28	4.21	4.15	4.10	4.06	4.00	3.94	3.87	3.84	3.81	3.77	3.74	3.70	3.67
7	5.59	4.74	4.35	4.12	3.97	3.87	3.79	3.73	3.68	3.64	3.57	3.51	3.44	3.41	3.38	3.34	3.30	3.27	3.23
8	5.32	4.46	4.07	3.84	3.69	3.58	3.50	3.44	3.39	3.35	3.28	3.22	3.15	3.12	3.08	3.04	3.01	2.97	2.93
9	5.12	4.26	3.86	3.63	3.48	3.37	3.29	3.23	3.18	3.14	3.07	3.01	2.94	2.90	2.86	2.83	2.79	2.75	2.71
10	4.96	4.10	3.71	3.48	3.33	3.22	3.14	3.07	3.02	2.98	2.91	2.85	2.77	2.74	2.70	2.66	2.62	2.58	2.54
11	4.84	3.98	3.59	3.36	3.20	3.09	3.01	2.95	2.90	2.85	2.79	2.72	2.65	2.61	2.57	2.53	2.49	2.45	2.40
12	4.75	3.89	3.49	3.26	3.11	3.00	2.91	2.85	2.80	2.75	2.69	2.62	2.54	2.51	2.47	2.43	2.38	2.34	2.30
13	4.67	3.81	3.41	3.18	3.03	2.92	2.83	2.77	2.71	2.67	2.60	2.53	2.46	2.42	2.38	2.34	2.30	2.25	2.21
14	4.60	3.74	3.34	3.11	2.96	2.85	2.76	2.70	2.65	2.60	2.53	2.46	2.39	2.35	2.31	2.27	2.22	2.18	2.13
15	4.54	3.68	3.29	3.06	2.90	2.79	2.71	2.64	2.59	2.54	2.48	2.40	2.33	2.29	2.25	2.20	2.16	2.11	2.07
16	4.49	3.63	3.24	3.01	2.85	2.74	2.66	2.59	2.54	2.49	2.42	2.35	2.28	2.24	2.19	2.15	2.11	2.06	2.01
17	4.45	3.59	3.20	2.96	2.81	2.70	2.61	2.55	2.49	2.45	2.38	2.31	2.23	2.19	2.15	2.10	2.06	2.01	1.96
18	4.41	3.55	3.16	2.93	2.77	2.66	2.58	2.51	2.46	2.41	2.34	2.27	2.19	2.15	2.11	2.06	2.02	1.97	1.92
19	4.38	3.52	3.13	2.90	2.74	2.63	2.54	2.48	2.42	2.38	2.31	2.23	2.16	2.11	2.07	2.03	1.98	1.93	1.88
20	4.35	3.49	3.10	2.87	2.71	2.60	2.51	2.45	2.39	2.35	2.28	2.20	2.12	2.08	2.04	1.99	1.95	1.90	1.84
21	4.32	3.47	3.07	2.84	2.68	2.57	2.49	2.42	2.37	2.32	2.25	2.18	2.10	2.05	2.01	1.96	1.92	1.87	1.81
22	4.30	3.44	3.05	2.82	2.66	2.55	2.46	2.40	2.34	2.30	2.23	2.15	2.07	2.03	1.98	1.94	1.89	1.84	1.78
23	4.28	3.42	3.03	2.80	2.64	2.53	2.44	2.37	2.32	2.27	2.20	2.13	2.05	2.01	1.96	1.91	1.86	1.81	1.76
24	4.26	3.40	3.01	2.78	2.62	2.51	2.42	2.36	2.30	2.25	2.18	2.11	2.03	1.98	1.94	1.89	1.84	1.79	1.73
25	4.24	3.39	2.99	2.76	2.60	2.49	2.40	2.34	2.28	2.24	2.16	2.09	2.01	1.96	1.92	1.87	1.82	1.77	1.71
30	4.17	3.32	2.92	2.69	2.53	2.42	2.33	2.27	2.21	2.16	2.09	2.01	1.93	1.89	1.84	1.79	1.74	1.68	1.62
40	4.08	3.23	2.84	2.61	2.45	2.34	2.25	2.18	2.12	2.08	2.00	1.92	1.84	1.79	1.74	1.69	1.64	1.58	1.51
60	4.00	3.15	2.76	2.53	2.37	2.25	2.17	2.10	2.04	1.99	1.92	1.84	1.75	1.70	1.65	1.59	1.53	1.47	1.39
120	3.92	3.07	2.68	2.45	2.29	2.18	2.09	2.02	1.96	1.91	1.83	1.75	1.66	1.61	1.55	1.50	1.43	1.35	1.25
∞	3.84	3.00	2.60	2.37	2.21	2.10	2.01	1.94	1.88	1.83	1.75	1.67	1.57	1.52	1.46	1.39	1.32	1.22	1.00

*This table is reproduced from M. Merrington and C. M. Thompson, "Tables of percentage points of the inverted beta (F) distribution," *Biometrika*, Vol. 33, 1943 (also in *Biometrika Tables for Statisticians*, Vol. I, Table 18), by permission of the *Biometrika* trustees.

TABLE VI (b) *F* RATIOS, $\alpha = 0.01$*

d.f. for Denominator	Degrees of Freedom for Numerator 1	2	3	4	5	6	7	8	9	10	12	15	20	24	30	40	60	120	∞
1	4,052	5,000	5,403	5,625	5,764	5,859	5,928	5,982	6,023	6,056	6,106	6,157	6,209	6,235	6,261	6,287	6,313	6,339	6,366
2	98.5	99.0	99.2	99.2	99.3	99.3	99.4	99.4	99.4	99.4	99.4	99.4	99.4	99.5	99.5	99.5	99.5	99.5	99.5
3	34.1	30.8	29.5	28.7	28.2	27.9	27.7	27.5	27.3	27.2	27.1	26.9	26.7	26.6	26.5	26.4	26.3	26.2	26.1
4	21.2	18.0	16.7	16.0	15.5	15.2	15.0	14.8	14.7	14.5	14.4	14.2	14.0	13.9	13.8	13.7	13.7	13.6	13.5
5	16.3	13.3	12.1	11.4	11.0	10.7	10.5	10.3	10.2	10.1	9.89	9.72	9.55	9.47	9.38	9.29	9.20	9.11	9.02
6	13.7	10.9	9.78	9.15	8.75	8.47	8.26	8.10	7.98	7.87	7.72	7.56	7.40	7.31	7.23	7.14	7.06	6.97	6.88
7	12.2	9.55	8.45	7.85	7.46	7.19	6.99	6.84	6.72	6.62	6.47	6.31	6.16	6.07	5.99	5.91	5.82	5.74	5.65
8	11.3	8.65	7.59	7.01	6.63	6.37	6.18	6.03	5.91	5.81	5.67	5.52	5.36	5.28	5.20	5.12	5.03	4.95	4.86
9	10.6	8.02	6.99	6.42	6.06	5.80	5.61	5.47	5.35	5.26	5.11	4.96	4.81	4.73	4.65	4.57	4.48	4.40	4.31
10	10.0	7.56	6.55	5.99	5.64	5.39	5.20	5.06	4.94	4.85	4.71	4.56	4.41	4.33	4.25	4.17	4.08	4.00	3.91
11	9.65	7.21	6.22	5.67	5.32	5.07	4.89	4.74	4.63	4.54	4.40	4.25	4.10	4.02	3.94	3.86	3.78	3.69	3.60
12	9.33	6.93	5.95	5.41	5.06	4.82	4.64	4.50	4.39	4.30	4.16	4.01	3.86	3.78	3.70	3.62	3.54	3.45	3.36
13	9.07	6.70	5.74	5.21	4.86	4.62	4.44	4.30	4.19	4.10	3.96	3.82	3.66	3.59	3.51	3.43	3.34	3.25	3.17
14	8.86	6.51	5.56	5.04	4.70	4.46	4.28	4.14	4.03	3.94	3.80	3.66	3.51	3.43	3.35	3.27	3.18	3.09	3.00
15	8.68	6.36	5.42	4.89	4.56	4.32	4.14	4.00	3.89	3.80	3.67	3.52	3.37	3.29	3.21	3.13	3.05	2.96	2.87
16	8.53	6.23	5.29	4.77	4.44	4.20	4.03	3.89	3.78	3.69	3.55	3.41	3.26	3.18	3.10	3.02	2.93	2.84	2.75
17	8.40	6.11	5.19	4.67	4.34	4.10	3.93	3.79	3.68	3.59	3.46	3.31	3.16	3.08	3.00	2.92	2.83	2.75	2.65
18	8.29	6.01	5.09	4.58	4.25	4.01	3.84	3.71	3.60	3.51	3.37	3.23	3.08	3.00	2.92	2.84	2.75	2.66	2.57
19	8.19	5.93	5.01	4.50	4.17	3.94	3.77	3.63	3.52	3.43	3.30	3.15	3.00	2.92	2.84	2.76	2.67	2.58	2.49
20	8.10	5.85	4.94	4.43	4.10	3.87	3.70	3.56	3.46	3.37	3.23	3.09	2.94	2.86	2.78	2.69	2.61	2.52	2.42
21	8.02	5.78	4.87	4.37	4.04	3.81	3.64	3.51	3.40	3.31	3.17	3.03	2.88	2.80	2.72	2.64	2.55	2.46	2.36
22	7.95	5.72	4.82	4.31	3.99	3.76	3.59	3.45	3.35	3.26	3.12	2.98	2.83	2.75	2.67	2.58	2.50	2.40	2.31
23	7.88	5.66	4.76	4.26	3.94	3.71	3.54	3.41	3.30	3.21	3.07	2.93	2.78	2.70	2.62	2.54	2.45	2.35	2.26
24	7.82	5.61	4.72	4.22	3.90	3.67	3.50	3.36	3.26	3.17	3.03	2.89	2.74	2.66	2.58	2.49	2.40	2.31	2.21
25	7.77	5.57	4.68	4.18	3.86	3.63	3.46	3.32	3.22	3.13	2.99	2.85	2.70	2.62	2.53	2.45	2.36	2.27	2.17
30	7.56	5.39	4.51	4.02	3.70	3.47	3.30	3.17	3.07	2.98	2.84	2.70	2.55	2.47	2.39	2.30	2.21	2.11	2.01
40	7.31	5.18	4.31	3.83	3.51	3.29	3.12	2.99	2.89	2.80	2.66	2.52	2.37	2.29	2.20	2.11	2.02	1.92	1.80
60	7.08	4.98	4.13	3.65	3.34	3.12	2.95	2.82	2.72	2.63	2.50	2.35	2.20	2.12	2.03	1.94	1.84	1.73	1.60
120	6.85	4.79	3.95	3.48	3.17	2.96	2.79	2.66	2.56	2.47	2.34	2.19	2.03	1.95	1.86	1.76	1.66	1.53	1.38
∞	6.63	4.61	3.78	3.32	3.02	2.80	2.64	2.51	2.41	2.32	2.18	2.04	1.88	1.79	1.70	1.59	1.47	1.32	1.00

*This table is reproduced from M. Merrington and C. M. Thompson, "Tables of percentage points of the inverted beta (*F*) distribution," *Biometrika*, Vol. 33, 1943 (also in *Biometrika Tables for Statisticians*, Vol. I, Table 18), by permission of the *Biometrika* trustees.

TABLE VII CRITICAL VALUES OF r*

d.f.	$r_{.05}$	$r_{.01}$
1	0.997	1.000
2	0.950	0.990
3	0.878	0.959
4	0.811	0.917
5	0.754	0.874
6	0.707	0.834
7	0.666	0.798
8	0.632	0.765
9	0.602	0.735
10	0.576	0.708
11	0.553	0.684
12	0.532	0.661
13	0.514	0.641
14	0.497	0.623
15	0.482	0.606
16	0.468	0.590
17	0.456	0.575
18	0.444	0.561
19	0.433	0.549
20	0.423	0.537
21	0.413	0.526
22	0.404	0.515
23	0.396	0.505
24	0.388	0.496
25	0.381	0.487
26	0.374	0.478
27	0.367	0.470
28	0.361	0.463
29	0.355	0.456
30	0.349	0.449
35	0.325	0.418
40	0.304	0.393
45	0.288	0.372
50	0.273	0.354
60	0.250	0.325
70	0.232	0.302
80	0.217	0.283
90	0.205	0.267
100	0.195	0.254
125	0.174	0.228
150	0.159	0.208
200	0.138	0.181
300	0.113	0.148
400	0.098	0.128
500	0.088	0.115
1000	0.062	0.081

*Reprinted by permission from *Statistical Methods* by G. W. Snedecor and W. G. Cochran, sixth edition. (c) 1967 by Iowa State University Press, Ames, Iowa.

TABLE VIII PROCEDURES FOR STATISTICAL INFERENCE

(Study your problem. Doesn't one of these outlines fit? Use it.)

I Estimation

1. PROPORTIONS

$n =$

$p =$

$c.i.\ \% =$

$$\hat{\sigma}_p = \sqrt{\frac{p(1-p)}{n}}$$

$z =$

2. MEANS

$n =$

$\bar{x} =$

$s =$

$c.i.\ \% =$

$$\hat{\sigma}_{\bar{x}} = \frac{s}{\sqrt{n}}$$

$z =$

$p \pm z\hat{\sigma}_p =$ ← confidence intervals → $\bar{x} \pm z\hat{\sigma}_{\bar{x}}$

II Testing Hypotheses

1. PROPORTIONS (NORMAL DISTRIBUTION)

$H_0: \pi =$

$H_1: \pi(\neq, >, <,)\ (?)$

$n =$

$p =$

$\alpha =$

$\bar{p} = \pi$

$$\sigma_p = \sqrt{\frac{\pi(1-\pi)}{n}}$$

$z =$

$\pi \pm z\sigma_p =$

2. MEANS (NORMAL DISTRIBUTION)

$H_0: \mu =$

$H_1: \mu(\neq, >, <)\ (?)$

$n =$

$\bar{x} =$

$s =$

$\alpha =$

$\mu_{\bar{x}} = \mu$

$$\hat{\sigma}_{\bar{x}} = \frac{s}{\sqrt{n}}$$

$z =$ or $t =$

$\mu \pm z\hat{\sigma}_{\bar{x}}$ or $\mu \pm t\hat{\sigma}_{\bar{x}} =$

3. PROPORTIONS (CHI-SQUARE)

H_0: no significant difference

H_1: is significant difference

$\alpha =$ $r =$ $c =$

$\text{d.f.} = (r-1)(c-1)$

$\chi^2_e =$

$$\chi^2_a = \sum \frac{(a-e)^2}{e} \text{ (for d.f.} > 1)$$

$$\chi^2_a = \sum \frac{(|a-e|-0.5)^2}{e} \text{ (for d.f.} = 1)$$

II Testing Hypotheses (*cont.*)

4. MEANS (ANOVA)

H_0: no significant difference
H_1: is significant difference
$\alpha =$ $r =$ $c =$
n.d.f. $= c - 1 =$
d.d.f. $= c(r - 1) =$
$F_e =$

$$s^2 = \sum \frac{(x - \bar{x})^2}{r - 1}$$

$$\hat{\sigma}_{\text{within}} = \frac{s_1^2 + s_2^2 + \cdots + s_c^2}{c}$$

$$\hat{\mu}_{\bar{x}} = \frac{\bar{x}_1 + \bar{x}_2 + \cdots + \bar{x}_c}{c}$$

$$\hat{\sigma}^2_{\text{between}} = r\left[\frac{\sum (\bar{x} - \mu_{\bar{x}})^2}{c - 1}\right]$$

$$F_a = \frac{\hat{\sigma}^2_{\text{b}}}{\hat{\sigma}^2_{\text{w}}}$$

5. PAIRED VARIABLES (CORRELATION)

H_0: no significant relationship
H_1: is significant relationship
$n =$
$\alpha =$
d.f. $= n - 2 =$
$r_\alpha =$
table for x, y, x^2, y^2, xy

$$r = \frac{n(\Sigma xy) - (\Sigma x)(\Sigma y)}{\sqrt{n(\Sigma x^2) - (\Sigma x)^2}\sqrt{n(\Sigma y^2) - (\Sigma y)^2}}$$

III Regression Equations

$n =$
table for x, y, x^2, y^2, xy

$$a = \frac{(\Sigma y)(\Sigma x^2) - (\Sigma x)(\Sigma xy)}{n(\Sigma x^2) - (\Sigma x)^2}$$

$$b = \frac{n(\Sigma xy) - (\Sigma x)(\Sigma y)}{n(\Sigma x^2) - (\Sigma x)^2}$$

$$y = bx + a$$

Reasonable Results for Exercises

These results may differ slightly from yours, depending on many factors, such as precision, accuracy, rounding off, and use or nonuse of the continuity adjustment. Furthermore, they are usually approximations—from assuming normal distributions, using sample statistics for population parameters, or employing other expedients. But the time has come to realize that in practical problems there is seldom a one and only "right answer."

Students have convinced me that results for every problem in the exercises should be furnished so that they may know whether their own efforts have produced reasonable results or whether they should try again. I have endeavored to do this and hope that *my* results turn out to be reasonable.

CHAPTER 1 (EXERCISES, page 4)

Possible Key Words or Phrases

1. Personal evaluation.
2. Good and bad chance?
3. Directed effort.
4. Cause and effect.
5. Is chance a cause?
6. Unimaginable?
7. Force implies effect.
8. Ever a single cause?
9. Is time a force?
10. Hammer and thumb.
11. Things are related.
12. Just what, not why.
13. Similar results.
14. Sampling.
15. Yes.

CHAPTER 2 (EXERCISES, page 12)

Possible Key Words or Phrases

1. People.
2. Example.
3. Bullets and bees.
4. Cross section.

5. Slanted.
6. Too regular.
7. Distortion by extremes.
8. Human error.
9. Closer scrutiny.
10. Equally likely.
11. Random numbers?
12. See # 4.
13. Numbered tags?
14. Population changes.
15. Constant chance.
16. Nausea.
17. $\frac{26}{52} \cdot \frac{25}{51} \cdot \frac{24}{50} \cdot \frac{23}{49} \cdot \frac{22}{48} = ?$
18. Census, counts.
19. Infinite populations.
20. Units of measurement.
21. In between.

CHAPTER 3 (EXERCISES, page 21)

Possible Key Words or Reasonable Numerical Responses

1. 0.000 through 1.000, inclusive.
2. Extremes.
3. A guess, or the result of sampling.
4. $P_{(\text{rolling a 2 with a die})} = 0.167.$
5. $P_{(\text{flipping a head, one coin})} = 0.500.$
   $P_{(\text{flipping not a head})} = 0.500.$
6. $P_{(\text{drawing ace of spades})} = 0.019.$
   $P_{(\text{not drawing ace of spades})} = 0.981.$
7. $P_{(\text{head or tail, one coin})} = 1.000.$
   $P_{(\text{neither head nor tail})} = 0.000.$
8. $P_{(\text{rolling die, 2 or 3})} = 0.334.$
9. $P_{(\text{ace of hearts or king of hearts})} = 2(0.019) = 0.038.$
10. Sample.
11. Sample.
12. $P_{(\text{underweight can contents})} = 0.004.$
13. Better methods.
14. Can you make it stand that way?
15. Just chance?
16. Never.

17. Numerical probability of being wrong.
18. Weather.
19. Only two possible outcomes on a single try.
20. Any item equally likely.

CHAPTER 4 (EXERCISES, page 30)

1. Heads or tails; 1, 2, 3, 4, 5, or 6 with a single die. One side out with a pool ball?
2. $P_{(\text{first pin})} = 0.008$; $P_{(\text{second pin})} = 0.055$; $P_{(\text{third pin})} = 0.164$; $P_{(\text{fourth pin})} = 0.273$; $P_{(\text{fifth pin})} = 0.273$; $P_{(\text{sixth pin})} = 0.164$; $P_{(\text{seventh pin})} = 0.055$; $P_{(\text{eighth pin})} = 0.008$.
3.

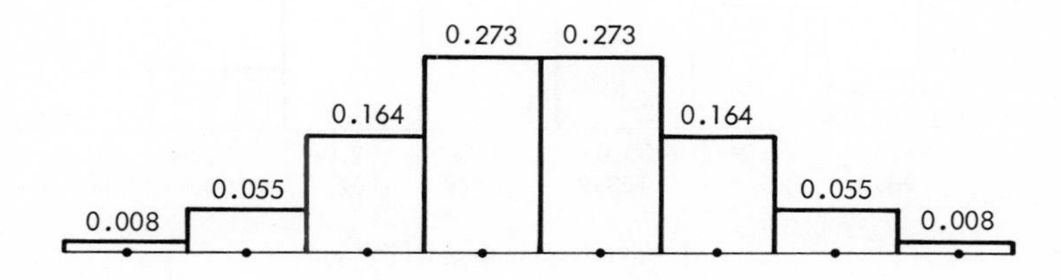

Probability distribution for hitting each of the eight pins in the seventh row

4. A central vertical line leaves one side the mirror image of the other.
5. $P_{(\text{3 heads, 9 coins})} = 0.164$.
6. $P_{(\text{3 heads, one coin, 9 tries})} = 0.164$.
7. $P_{(\text{more than 3 heads, 9 tries})} = 0.746$.
8. $P_{(\text{3 heads or less, 9 tries})} = 0.254$.
9. $P_{(\text{less than 3 heads, 9 coins})} = 0.090$.
10. $P_{(\text{3, 4, or 5 heads, 8 coins})} = 0.711$.
11. $P_{(\text{4 red beads out of 10})} = 0.251$.
12. $P_{(\text{one defective brush in 15})} = 0.366$.
13. $P_{(\text{no hearts, 13 tries with replacement})} = 0.032$.
14. $P_{(\text{all 10 claim relief})} = 0.107$.
15. $P_{(\text{1 underweight in 10})} = 0.315$.
16. Find a larger binomial probability table.
17. The 80% goes with the hydralazine hydrochloride in individual pills. There is no information about how this varies.
18. The probability of a head remains the same all the time. The "law of averages" says that, in the long run, heads will turn up about half the time.
19. Yes, making up problems of your own is an excellent way to learn.

CHAPTER 5 (EXERCISES, page 40)

1.

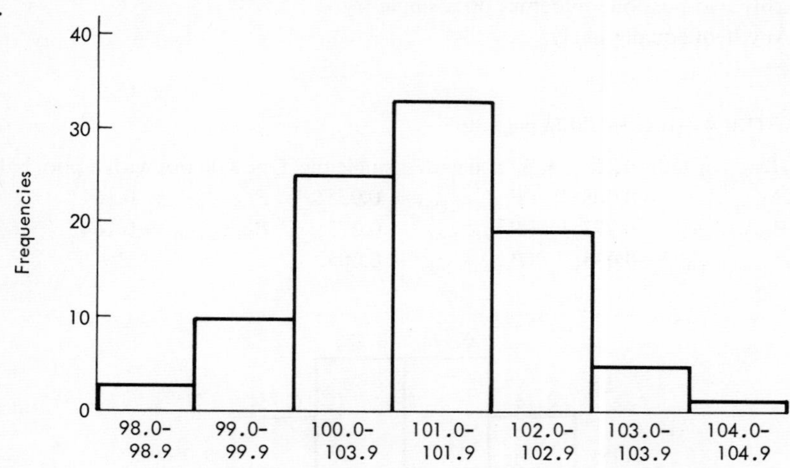

2. Given the standard deviation in inches, gallons, or microvolts, the standard normal distribution shows the proportions of the total area under the curve between the mean and any specific number of these deviations.
3. (a) $P_{(\mu \pm 1.00\sigma)} = 0.6826$ or 68%.
   (b) $P_{(\mu \pm 2.00\sigma)} = 0.9544$ or 95%.
   (c) $P_{(\mu \pm 3.00\sigma)} = 0.9974$ or 99^{+}%.

   The rest of our probabilities we shall show to three decimal places.
4. (a) $P_{(\text{between } z=0 \text{ and } z=1.50)} = 0.433$.
   (b) $P_{(\text{between } z=0 \text{ and } z=2.31)} = 0.490$.
5. (a) $P_{(\text{between } z=0 \text{ and } z=-1.50)} = 0.433$.
   (b) $P_{(\text{between } z=0 \text{ and } z=-2.97)} = 0.498$.
6. (a) Part of area between $z = 1.5$ and $z = -1.5$ is 0.866.
   (b) Part of area between $z = -1.78$ and $z = 2.70$ is 0.959.
7. (a) $P_{(\text{beyond } z=1.96, \text{ farther from the mean than } z=1.96)} = 0.025$.
   (b) $P_{(\text{beyond } z=-2.33 \text{ or } z=2.33)} = 0.020$.
8. $z_{0.3485} = 1.03, -z_{0.4778} = -2.01$.
9. $z_{0.4900} = 2.33$.
10. (a) $P_{(\text{under } 40)} = 0.663$.
    (b) 27 to 43.
11. $P_{(\text{just 3 out of 10 will die})} = 0.010$.
12. $\mu \pm 3\sigma = 80$ mm to 170 mm. These are heights in millimeters of mercury in the pressure tube.
13. $P_{(\text{below } 270)} = 0.138$.

14. $P_{(\text{less than 900 hours})} = 0.159$.
15. They are probably not related.
16. 4 to 17 years.
17. You don't have the average number of absent employees, which you must have in order to calculate.
18. $P_{(2 \text{ or more strikes in 7 drillings})} = 0.423$.

CHAPTER 6 (EXERCISES, page 61)

1. $\sum x = 214$, to the nearest unit.
2. $\bar{x} = 21.4$, one decimal place to calculate with.
3. -2.4, $+1.6$, $+0.6$, -2.4, $+1.6$, $+2.6$, -4.4, $+0.6$, $+0.6$, $+1.6$. Add or subtract with one more decimal place than given, using $\bar{x} = 21.4$.
4. The sum of the squares is 46.40.
5. $s = 2.272$.
6. 14 to 29 onions per bunch.
7. $\sum x = 19.9$ grams, to the nearest tenth of a gram.
8. $\bar{x} = 1.99$. Three significant figures in 19.9, so keep three.
9. But we are going to subtract measurements to the nearest tenth, so round off 1.99 to 2.0. $\sum(x - \bar{x})^2 = 0.69$ grams.
10. In the text exercise we got 0.7. Here we get 6.9. Let's see how much difference it makes.
11. $s = \sqrt{\frac{\sum (x - \bar{x})^2}{n - 1}} = \sqrt{\frac{0.7}{9}} = \sqrt{0.077} = 0.28$

    $= \sqrt{\frac{0.69}{9}} = \sqrt{0.077} = 0.28$

    Both the same; round to 0.3.
12. 1.1 to 2.9 grams.
13. Sample too large.
14. $P_{(\text{over 20 tremors per second})} = 0.002$.
15. Sample too large.
16. We don't know what you got, but wasn't it close to rectangular?
17. The two significant figures for 63 miles are least accurate, so how about 63/208,000 to give us 0.030%?
18. In the 69.5 to 70.5°F interval.
19. Could be 97.76 to 97.85°F.
20. The areas differ by 41 sq ft. $18 \times 23 = 414$ sq ft. Wouldn't it be safe to round off to the given number of significant figures (410 sq ft)?

CHAPTER 7 (EXERCISES, page 73)

1. The individual binomial probability must be constant, and, for our tables, the sample size must be less than 16.

2. Both $n\pi$ and $n(1 - \pi)$ should be greater than 5. Your binomial tables are no help because n is greater than 15. You need n, π, and x.
3. The weights in grams of 100 crows would not be binomial. It could even be that the results are all different; they are not just *either, or.*
4. Use Tchebycheff's theorem when you don't know that the distribution is normal. You need only the standard deviation (σ) and the number of standard deviations (k) that the value in question (x) is from the mean (μ).
5. $P_{(\text{red 7 times out of 15})} = 0.196$.
6. $P_{(\text{4 out of 14 elms will die})} = 0.155$.
7. $P_{(\text{less than 20 support mayor})} = 0.001$.
8. $P_{(\text{all listening to instructor})} = 0.005$.
9. $P_{(\text{8 reds in 15 spins})} = 0.176$.
10. $\bar{x} \pm 1s = 99.985$ to $100.015°C$.
    $\bar{x} \pm 2s = 99.970$ to $100.030°C$.
    $\bar{x} \pm 3s = 99.955$ to $100.045°C$.
11. $P_{(\text{grass blade exceeds 2.9 inches})} = 0.000$.
12. At least $\frac{15}{16}$ within 4σ of μ.
13. At least $\frac{3}{4}$ bluejay beaks between 1.90 and 2.50 cm.
14. $\bar{x} = 3.8$ cm.
15. $s = 1.28$.
16. At least $\frac{3}{4}$ of motor lives between 850 and 1350 hours. Continuity adjustment would be insignificant.

CHAPTER 8 (EXERCISES, page 89)

1. Both $n\pi$ and $n(1 - \pi)$ should be greater than 5.
2. Since theoretically the mean proportion, $\bar{p}$, of all samples the same size is equal to π we can use $\sigma_p = \sqrt{\frac{(\pi 1 - \pi)}{n}}$ for $\sigma_p = \sqrt{\frac{\bar{p}(1 - \bar{p})}{n}}$
3. $P_{(\text{between 0.70 and 0.80 totally dependent})} = 1.000$.
4. $P_{(\text{over 0.40 own two cars})} = 0.015$.
5. $P_{(\text{between 0.045 and 0.055 weed seed})} = 0.978$.
6. $P_{(\text{8 out of 300 will break a leg})} = 0.116$.
7. Too far out; $p = 0.69$ gives $z = 8.26$. Cards not properly shuffled.
8. $P_{(\text{more than 0.20 will not live to 60 years})} = 1.000$.
9. $P_{(\text{a randomly selected person is an alcoholic})} = 0.005$.
10. The aqueous solution proportion does not tell you what proportion of the items in the population have a certain characteristic.
11. $P_{(\text{of hitting over 0.390 in first 50})} = 0.164$.
12. $P_{(\text{0.02 or more will be undersize})} = 0.719$.
13. $P_{(\text{more than 0.55})} = 0.159$.
14. See # 10.

15. $P_{(0.50 \text{ or more vote for Jones})} = 0.023$.
16. Should stop it when p gets up to 0.002.
17. $P_{(\text{more than } 0.24 \text{ births are unwanted})} = 0.062$.

CHAPTER 9 (EXERCISES, page 99)

1. Distribution of sample means is approximately normal.
2. Should have sample size 30 or more.
3. Sample size should be less than 5% of the population.
4. $\sigma_{\bar{x}} = \dfrac{\sigma_x}{\sqrt{n}}$.
5. $P_{(\text{mean weight between } 195 \text{ and } 200 \text{ pounds})} = 0.477$.
6. $P_{(\text{mean between } 6.98 \text{ and } 7.02 \text{ oz})} = 0.205$.
7. $P_{(\text{over } 10.6 \text{ cm})} = 0.006$.
8. $P_{(10 \text{ mm, average lowering of blood pressure})} = 0.008$.
9. $P_{(\text{less than } 0.60 \text{ are Republican})} = 0.203$.
10. Impossible—no σ.
11. At least $\frac{3}{4}$ of the textbooks between $9 and $11.
12. Impossible; no standard deviation again.
13. $P_{(\text{at least } 6 \text{ right out of } 10)} = 0.377$.
    $P_{(\text{at least } 60 \text{ right out of } 100)} = 0.029$.
14. The average number of turns in 100 different feet would be 46.6 to 47.4.
15. Set machine at 16.33 oz.
16. $P_{(\text{sample mean less than } 49{,}000 \text{ mi})} = 0.000$.
    $P_{(\text{single tire less than } 49{,}000 \text{ mi})} = 0.115$.

CHAPTER 10 (EXERCISES, page 109)

1. The sample proportion tells you that the exact population proportion probably lies within $p \pm 3\hat{\sigma}_p$ of that proportion.
2. If the sample size is 30 or more, the distribution of the sample means will be approximately normal.
3. $\bar{x}$, s, and n. You'll need sample size, mean, standard deviation, c. i. percent.
4. It says with a specified percent of confidence that the parameter lies in this interval.
5. Round off toward more confidence: 0.18 to 0.26 for unwanted childbirths.
6. Bass percentage, 0.26 to 0.32.
7. Prescription estimate, 0.20 to 0.40.
8. Hospitalization estimate, 0.501 to 0.659.
9. At least $\frac{3}{4}$ of the test scores between 55 and 95.
10. $P_{(\text{exactly } 5 \text{ out of } 100 \text{ live})} = 0.182$.
11. Proportion of objecting wives, 0.66 to 0.84.

12. (a) When both $n\pi$ and $n(1 - \pi)$ are greater than 5.
    (b) Same as above.
    (c) When $n \geq 30$.
13. Effective insect killing between 62 and 78 per hundred.
14. Suicide rate 229 to 293 per 1,000,000.

CHAPTER 11 (EXERCISES, page 119)

1. You need n, $\bar{x}$, s, and also the confidence interval percent.
2. As the sample increases above 30, s rapidly gets closer to σ.
3. $n = 50$. Total mileage = 1000. Population is every mile for car and driver. $\sigma > \hat{\sigma}_{\bar{x}}$.
4. 95% of the pumpkinseeds are between 147 and 153 mm.
5. Probably 8.45 to 8.60 oz is the safest 99% estimate.
6. Well, 158,000 to 168,000 cars.
7. No standard deviation; can't do.
8. Mean time at 95% confidence; 17.3 to 18.7 minutes.
9. Mean parametric number of loadings between 417 and 429.
10. Weekly cost of janitorial service: $32.79 to $41.67.
11. $P_{(\text{accepting ledger})} = 0.012$.
12. $P_{(\text{waiting 20 minutes or more})} = 0.067$.
13. $P_{(\text{just 7 will grow})} = 0.009$. The most probable number to grow is 11 or 12.
14. No standard deviation again.
15. Average children's hours at TV: 14 to 16 hours. This applies to that Delaware town.
16. $P_{(\text{less than one pound})} = 0.023$.
17. $P_{(n=30,\ \bar{x}<1\text{ lb})} = 0.000$.
18. $P_{(\bar{x}<1.00)} = 0.500$.
19. Try to get $\mu_{\bar{x}} = 1.003$ lb.
20. Using t, 95% interval is, safely: 7928 to 8072 lbs, for $n = 10$.

CHAPTER 12 (EXERCISES, page 135)

1. $H_0: \pi = 0.2$; $\quad H_1: \pi \neq 0.2$.
2. $H_0: \pi = 0.55$; $\quad H_1: \pi \neq 0.55$.
3. $H_0: \mu = 30{,}000$ miles; $\quad H_1: \mu < 30{,}000$ miles.
4. $H_0: \pi = 0.01$; $\quad H_1: \pi > 0.01$.
5. $p = 0.13$, between 0.10 and 0.30; do not reject H_0: 3 out of 15 still holds.
6. $p = 0.57$, between 0.52 and 0.58; do not reject: Jones's 0.55 still holds.
7. $p = 0.05$, between 0.024 and 0.116; do not reject: dean's information probably correct.
8. $p = 0.60$, below 0.603; reject H_0: Forestry Service seems to be wrong. Too close, though.

9. Check export records.
10. 99% c.i. = safely, 0.39 to 0.43 have type A.
11. $p = 0.031$, below 0.049; do not reject: processor's claim appears to be correct.
12. $P_{(\text{high-school juniors} < 80)} = 0.062$.
13. $p = 0.38$, above 0.33; do not reject: pharmaceutical claim seems to be correct.
14. $p = 0.80$, below 0.93; do not reject: salve seems to be no better.
15. $p = 0.56$, above 0.54; reject H_0: medium *has* it.
16. $p = 0.17$, below 0.194; reject H_0: commentator's claim appears to be wrong.
17. $p = 0.008$, between 0.000 and 0.010; do not reject: sociologist's claim holds.
18. Not clear. But if 63,472 lb/sq in. is $\bar{x}$, then 99% c.i. = 58,970 to 67,970 sq in.

CHAPTER 13 (EXERCISES, page 154)

1. Apparently Newton considered hypotheses to be intuitive or theoretical.
2. $H_0 : \mu = 46$ sec; $H_1 : \mu < 46$ sec.
3. $H_0 : \mu = \$8000$; $H_1 : \mu \neq \$8000$. Agency *could*, though, be saying "at least \$8000," making test one-sided.
4. $\bar{x} = 9.3$ sec, below 9.86; reject H_0: average fuse burning is not 10 seconds.
5. $\bar{x} = 325°$C, below 326.2°C; reject H_0: melting point of lead does not seem to be 327°C.
6. $\bar{x} = 215$, above 204; reject H_0: new machine is better.
7. $\bar{x} = \$81.23$, above \$77.78; reject H_0: furniture dealer's claim is too low.
8. Do not reject: sample mean is over 8 oz and cannot suggest that $\mu < 8$ oz.
9. $\bar{x} = 105$, below 106.3; reject H_0: breaking strength is less than claimed.
10. No; standard deviation is missing.
11. $\bar{x} = 99$, less than 99.12; reject H_0: the theoretical $\mu = 100$ seems to be wrong.
12. Practically all rods between 1.9973 to 2.0027 inches.
13. $p = 0.70$ is above 0.625; do not reject; radioactive iodine probably 75% effective.
14. 95% c.i. = 17.96 to 19.24 cm.
15. Nothing about a normal distribution, but Tchebycheff's theorem yields a probability of at least $\frac{8}{9}$ or 0.889.
16. 95% c.i. = 88.5 to 93.5 cm.
17. Can you find anyone who agrees with your results on this one?
18. $p = 0.72$, less than 0.727; reject H_0: fewer than 75% respond favorably. Pretty close, though.

CHAPTER 14 (EXERCISES, page 167)

1. $\chi^2_e = 6.635$, $\chi^2_a = 3.17$. Do not reject H_0: geneticist's claim for proportion of white-eyed fruit flies holds.
2. $p = 0.27$, between 0.227 and 0.273. Do not reject H_0 (but just barely, using normal distribution approach).

3. $\chi^2_e = 6.635$, $\chi^2_a = 2.41$. Do not reject H_0: no significant difference between proportion of married and single smokers.
4. $\chi^2_{el} = 6.635$, $\chi^2_a = 7.46$. Reject H_0: drug is probably effective.
5. $\chi^2_e = 6.635$, $\chi^2_a = 6.48$. Do not reject H_0: no significant difference in detergent preferences.
6. $\chi^2_e = 3.841$, $\chi^2_a = 3.91$. Reject H_0: significant difference in soft-drink vending machines.
7. Not ready for this one yet.
8. $\chi^2_e = 5.991$, $\chi^2_a > 5.991$. Reject H_0: there is a relationship between hair and eye color.
9. $\chi^2_e = 7.815$, $\chi^2_a > \chi^2_e$. Reject H_0: significant relationship between the numbers in the rows of different columns.
10. $\chi^2_e = 9.488$, $\chi^2_a < \chi^2_e$ (nine cells, only 1.5 is not a fraction of 1.0—no need to compute and add). Do not reject H_0: seems to be no relationship between height and self-control.
11. $\chi^2_e = 13.277$, $\chi^2_a = 7.47$. Do not reject H_0: no significant difference between soft-drink preferences in the different cities.
12. $\chi^2_e = 9.210$, $\chi^2_a < \chi^2_e$ (only two of the six cells are slightly more than 1—not necessary to compute completely). Do not reject H_0: no significant difference in effectiveness among the three drugs.
13. $\chi^2_e = 5.991$, $\chi^2_a < \chi^2_e$ (only one of the six is slightly more than 1). Do not reject: seems to be no significant relationship between number of responses and degrees of respondent.
14. $\chi^2_e = 16.812$, $\chi^2_a > \chi^2_e$ (this is obvious by the fourth or fifth step). Reject H_0: there is a significant relationship between number of children and income.
15. $\chi^2_e = 12.592$, $\chi^2_a > \chi^2_e$ (obvious very soon in computing χ^2_a). Reject H_0: there is a relationship between selection of female mate and listed characteristics.

CHAPTER 15 (EXERCISES, page 186)

1. $F_e = 7.71$, $F_a = 1.83$. Do not reject H_0: no apparent significant difference in route times.
2. $F_e = 5.32$, $F_a = 1.58$. Do not reject H_0: no significant difference in tensile strengths.
3. $F_e = 5.29$, $F_a = 9.68$. Reject H_0: seems to be a significant difference in pig feeds.
4. $F_e = 3.24$, $F_a < F_e$ (a fractional part of 1). Do not reject H_0: seems to be no significant difference in machine operators.
5. $F_e = 8.29$, $F_a = 321$. Reject H_0: test seems to differentiate between the two groups of students.
6. $F_e = 4.07$, $F_a < 1$. Do not reject H_0: there is no significant difference in gas mileage.

7. $\chi_e^2 = 12.592$ (still using $\alpha = 0.05$), $\chi_a^2 < 1$. Do not reject H_0: no significant difference between effectiveness of different brands of gas in different makes of cars.
8. $p = 0.65$, less than 0.73. Reject H_0: guitarist probably overclaimed. Different conclusion if two-sided test.
9. You have μ; you can construct the $\mu \pm 2.33\sigma_{\bar{x}}$ interval and *see* whether $\bar{x} = 36$ inches falls in it. It doesn't; it can't. If you insist: $P_{(\bar{x}=36 \text{ inches in } 98\% \text{ c.i.})} = 0.000$.
10. $F_e = 7.59$, $F_e = 5.3$. Do not reject H_0: no significant difference in durability of enamels.
11. $\chi_e^2 = 13.277$, χ_a^2 (must be less than 9). Do not reject H_0: number of cans in shelf frontage appears to make no significant difference.

CHAPTER 16 (EXERCISES, page 205)

In most of these correlation problems you are given presumably exact counts and only have to round off in division or in taking square roots. However, when you struggle with approximate numbers like heights in inches or temperatures in degrees centigrade, you could simplify some computations a bit by careful and thoughtful decisions about rounding off. But during some prolonged calculations it just may be easier not to wrestle with these computational decisions, particularly if you use a desk calculator; just round off final and intermediate results.

In the text, I used r_α for the minimum correlation coefficient. Here I have used r_e, instead, to conform with our use of χ_e^2 and F_e.

1. (a) +, (b) −, (c) +, (d) 0, (e) +, (f) +, (g) almost 0—linesmen do have generally higher numbers than men in the backfield, (h) 0, (i) +, (j) −.
2. These correlation coefficients are not common in the world of nature and chance. But: $100 at 6% per year shows a +1 correlation coefficient for principal plus interest and time. The height of a slanting roof shows a correlation of −1 with the distance from the ridgepole to eaves. There is apparently no correlation between the average daily price of cheese and the number of deaths from pneumonia.
3. $r_e = 0.997$, $r_a = 0.868$. Do not reject H_0: there appears to be no significant relationship between quiz grade and time spent doing the quiz.
4. $r_a = +1$, of course. $y = 1.07x - 1.57$.
5. $r_{e(\text{at } \alpha=0.05)} = 0.997$, $r_{e(\text{at } \alpha=0.01)} = 1.000$. $r_a = 0.979$. Reject H_0 at both levels: no significant relationship between fraternal twins and IQ's.
6. $r_e = 0.666$, $r_a = 0.778$. Reject H_0: there seems to be a significant relationship between students' heights and their fathers' heights. $y = 0.68x + 3.4$. If student's height is 73″, predict fathers' to be 72.2″.
7. $p = 0.20$, between 0.18 and 0.33. Do not reject H_0: 0.25 of women married two years or more claim unhappy marriage.

8. $F_e = 3.24$, $F_a = 2.06$. Do not reject H_0: no significant difference between machine operators.
9. $\chi_e^2 = 13.277$, $\chi_a^2 > \chi_e^2$. Reject H_0: sample shows a significant relationship between hair color and which of the three states is place of birth.
10. Estimate of average expenditure for noon lunch, \$1.75 to \$1.95.
11. $r = -0.65$. Negative correlation between Sz and precipitation. Not significant at either $\alpha = 0.05$ or $\alpha = 0.01$.
12. $r_e = 0.811$, $r_a = 0.840$. Reject H_0: significant relationship between amount of Concord and Catawba grapes is used.
13. $r_e = 0.878$, $r_a = 0.98$. Reject H_0: correlation between grades and total hours of weekly study is significant. $y = 4.7x + 0.6$.
14. $p = 80.1$, over 77.6. Reject H_0: pollution in river is more than claimed.
15. $r_a = 1$, $y = 0.98y + 37.2$.
16. 95% c.i. for $\bar{x}$'s of IQ's ($n = 2000$) is about 104 to 106.

Index